NEUROTRANSMISSIONS: ESSAYS ON PSYCHEDELICS FROM BREAKING CONVENTION

EDITED BY DAVE KING, DAVID LUKE, BEN SESSA, CAMERON ADAMS, AIMEE TOLLAN

First published by Strange Attractor Press 2015
© The Authors 2015

ISBN 978-1-907222-43-6

Strange Attractor Press
BM SAP, London, WC1N 3XX, UK
www.strangeattractor.co.uk

Printed and bound in Great Britain by TJ International Ltd. Padstow

Cover image by Blue Firth
Layout and design Emerald Mosley

DEDICATION

Dr Anna Waldstein, former director of Breaking Convention, for her hard work on Breaking Convention 2011 and 2013.

Sue King, for her support, love and enthusiasm. We'll see you somewhere downstream.

NEUROTRANSMISSIONS:
ESSAYS ON PSYCHEDELICS
FROM BREAKING CONVENTION

INTRODUCTION

Let me tell you of a strange thing to be observed in the psychedelic communities – a thing of which you are probably well aware already. It is this: the population of these communities is extraordinarily diverse. This is strange because in other parts of academia, there isn't nearly as much communication between disciplines as there should be. Immunology conferences tend to be attended by immunologists, while astrophysics conventions have very few immunologists among their delegates. That makes sense really, doesn't it? The academia of psychedelics is not like that. It's not at all specific. There are few other types of academic conference where one can observe a conversation happening between a Harvard professor, a Huichol mara'kame, a Rastafari ambassador, an ageing rock star, and a sweet old lady (whom may be easily imagined fitting in very nicely at the café down the road where they serve scones for afternoon tea). Indeed, the whole subject matter is so non-specific that even the oldest and wisest speakers at psychedelic conferences may struggle to explain what makes a psychedelic thing psychedelic. There is certainly something that connects the keynote talk by the professor of clinical pharmacology with the visionary art gallery next door, but it's not quite clear what it is. There are anthropologists and astronomers, chemists and choreographers, philosophers and phlebotomists, all of whom have somehow arrived at the same event despite coming from very different places.

After a little while you might notice something else a bit confusing: nobody seems to have trouble understanding anybody else, but they are all using different words. The psychiatrists are talking about *psychotomimetics* (at least, the older ones), the neuroscientists have charts of $5\text{-}HT_{2A}$ *receptor agonists,* the artists seem to prefer the term *visionary,* and some of the witches (yes, there are witches there too) are talking about *thaumaturgic* compounds. You overhear people talking about mysticomimetics, hallucinogens, deleriants, delusinogens, dissociatives, phantasticants and, of course, psychedelics. Not only that, but there are researchers talking about toads, pulsing light machines, herbs, breathing techniques, cacti, orgasms, drum circles, sweat lodges, mushrooms, and long spells spent in very dark, very quiet places, as if they were all talking about the same thing. Does anybody really know what they're talking about?

> *"If there is confusion in choosing a term to describe the class of drugs that we shall call the psychotomimetics, then there is chaos in agreeing upon a description of their effects."*   *(Shulgin, 1978)*

A BRIEF HISTORY OF VINES (AND OTHER SHORT STORIES)

Psychedelics have probably been used for thousands of years. It's difficult to prove that, of course, but we can make some educated guesses. Evans-Schultz & Hofmann (1979) estimated the first use of pituri among the Australian aborigines at about 40.000 years ago. Geometric patterns and suspiciously mushroom-like paintings in ancient caves have suggested, to some, indications of psychedelic states of consciousness in prehistoric man. Cannabis has been found in large quantities next to occupants of some very old graves, such as the 5,000-year-old Yanghai Tombs. The little that we do know of humankind's past relationship with psychedelics implies that these experiences were treated with reverence, and held spiritual implications. We know that the ancient Greeks celebrated the mysteries of Eleusis with a mysterious mind-altering drink known as *kykeon*, and that over a hundred of the hymns of the Rig-Veda praise a plant-drug of unknown origin, *soma* (good evidence is provided in Chapter 12 to suggest that it may have had similarities to ayahuasca). In these instances, the psychedelic was understood as sacrament.

What we might call 'contemporary Western society' was deprived of psychedelics for hundreds of years. We can say 'deprived' because, as we shall see in Chapter 14, there has been a history of ecstatic states of consciousness suffering from a degree of state suppression that is much older than the drug conventions of the United Nations. Though European cultures were not psychedelic for a very long time (we can call ancient Greece a psychedelic culture insofar as they had an approved, institutional place for the psychedelic state that was accessible to an annual pilgrimage of citizens, and we can say that, for instance, England during the middle ages was not a psychedelic culture in that there is no record of such use), certain cultures around the world have benefited from long-standing, unbroken relationships with psychedelic plants. These plants did not receive much recognition in Europe until the end of the Nineteenth Century, when researchers such as Louis Lewin and Havelock Ellis discovered peyote and started experimenting with it. Their discovery was not new, as various indigenous peoples of North America can tell you, but it was new for them.

Non-drug-induced visionary states of mind were better known, such as the prophetic states of the Old Testament, but developments in mid-Nineteenth Century French psychiatry sought to pathologise these visionary states, and by the end of the century hallucinations and ecstatic experience were associated with mental ill-health and instability. The word 'hallucination' was used as early as 1646 by the physician Thomas Browne, with a definition much like today's, but it was the French psychiatrist Jean-Étienne Esquirol who included the term in his diagnostic dictionary of mania in 1817, and in doing so coupled hallucination to mental disorder. The origin of the word can be traced to the Latin word *halucinari*, which referred to a sort of absentmindedness, or, in a translation of the Second Century writer Aulus Gellius, a "degenerate dreaminess". In 1845, Moreau de Tours wrote that "a hallucination is the most frequent symptom and the fundamental fact of delirium, mental illness and madness". The historian Mike Jay (2013) gives a marvellous account of all this in his essay 'Dreaming While Awake'; he notes that by the turn of the century not only had Moses, Socrates and Muhammad received retroactive diagnoses of epilepsy, hysteria and paranoia, and Joan of Arc with religious mania, but that the everyday public had begun to think of hallucination as a sign of madness. This was unfortunate, considering that hallucination is a common phenomenon among the sane. S. Weir Mitchell had a very difficult time getting soldiers to admit to experiencing phantom limb symptoms in 1872 – though the condition had been well recognised as early as Descartes.

Ben Sessa, an editor of this book and a director of Breaking Convention, suggests that psychedelic research falls into three epochs: the first being the early mescaline experiments, the second being the two decades of research that followed the discovery of LSD, and the third being the renaissance of research that is happening today (Sessa, 2012). At the beginning of the second epoch, early publications show confusion about how to categorise the effects of LSD and related compounds. The first categorical noun for the psychedelic plants had been proposed in 1912 by Lewin, who called them *phantastica*. In the late 1940s there were some researchers thinking in the same way, those who called the compounds "agents of phantasy"; others, in the line of the psychiatric tradition we have just discussed, called them "agents of psychosis". It was the latter that became the predominant assumption, as can be seen by the widespread use of the term *psychotomimetic* at the time.

Hoffer & Osmond wrote that "the first comprehensive clinical reports appeared in the European psychiatric literature. The experience was described as an exogenous or toxic reaction of the Bonhoeffer type. These clinical reports have not been surpassed." The Bonhoeffer reaction is a complex of mental symptoms resulting from a poisoning of the nervous system, which is more or less the reasoning of the psychiatric profession at the time: psychomimetics were drugs that caused a transient model psychosis through a poisoning effect (since psychosis was an illness, a drug that engenders it must be a poison). This was very exciting, because psychiatrists felt that they now had tools to understand the acute schizophrenic experience and offer routes of treatment (this led to the 'Toxin M' hypothesis, which proposed that an endogenous compound with activity similar to LSD must be present in the brains of schizophrenics; Osmond suggested that this might be adrenochrome, a metabolite of adrenaline). Hoffer & Osmond were particularly interested by the similarity between LSD and the toxic psychosis of *delirium tremens*, a syndrome of alcoholism produced upon withdrawal. If they were able to give alcoholics an artificial *delirium tremens*, without the cardiovascular risks, they might be able to scare the patients off the booze.

However, after some time, researchers began to realise that LSD and related compounds did not always produce a psychotic episode, and indeed in the cases where patients were helped it was not the psychotomimetic activity which was responsible. Although clinicians were trying hard to produce psychosis in their patients – with good intentions, I may add – many participants were having a completely different outcome. "In spite of our best efforts to produce a [psychotic] experience, some of our subjects escaped into a psychedelic experience" (Hoffer & Osmond, 1967). Thus, new terms entered the drug literature, including *psychedelic* (mind-manifesting), *psycherhexic* (mind-bursting forth), *psycheplastic* (mind-molding), *psychehormic* (mind-rousing) and other similar constructions. Ronald Sandison, who first brought LSD into the United Kingdom in 1952, was never attached to the model of psychotomimesis and proposed the term *psycholytic* (mind-releasing). Stanislav Grof, who developed a mind-altering breathwork technique following the global ban on psychedelic compounds, described the experiences as *holotropic* (moving towards wholeness).

Then the drugs ran away from the clinics and were swept up in the loving embrace of the 1960s anti-war counter culture. Little needs to be said about that period, because so much has been said about it already, but what is worth saying is that the word 'psychedelic' was plucked away from the intelligentsia and pinned forever onto colourful music posters and popular representations of the sexual revolution (a collection of such images has recently been edited by Hanson, Godtland & Krassner in their book *Psychedelic Sex*). In 1979, in the hope of finding a word that expressed the sincerity with which so many psychedelic users approached their experiences, and to avoid the hedonistic associations that 'psychedelic' had picked up, a group of researchers proposed the term *entheogen* (generating the inner divine). In Chapter 4, we see yet another addition to the psychedelic dictionary, *epilogen* (choice-enabling), and there is also the story of the *empathogens*, which is covered in Chapter 7.

There are at least 40 different terms so far that have been proposed for the psychedelics, and this number grows each year. This reflects a difficulty on our parts to describe an experience that is much too large for a single word, however ambitious. As Kluver wrote in 1928, upon taking these compounds "we enter a world beyond language, so it is hardly surprising that they may be difficult to name" (Shulgin, 1978). But that is not the only conclusion we can draw – we can also point out that which we noticed at the beginning of this introduction, that these compounds are used by a panoply of people, for different reasons, bringing different expectations, with different outcomes.

THE FRUITS OF NON-SPECIFICITY

This book is both interdisciplinary and inconsistent in its use of words. I hope the reasons for this are now clear. Here, in this volume, we reap the benefit of the non-specificity of psychedelic research.

The first chapters introduce psychedelic consciousness to the reader with a selection of papers on philosophy (Chapters 2 & 3), neuroscience (Chapters 1 & 5) and theoretical psychology (Chapter 4). Then, with a new understanding of current models of action, the reader will be guided through the clinical uses of psychedelic substances, with a particular focus on the psychedelic amphetamine MDMA (Chapters 7, 8 & 9). A detailed case study of personal growth through psychedelics is described in Chapter 6, and interpreted through Jungian psychological theory.

In following chapters we are given a chance to broaden our perspectives as we depart from the hospitals and the neuroscience departments and take a journey through space and time as we move through a history of psychedelics in anthropology (Chapter 11) learning about psychedelic cultures in New Guinea (Chapter 10), Central and South America (Chapter 13), Music Festivals (Chapters 15 & 16), and onwards to the ancient civilisations of India and Rome (Chapters 12 & 14). Sailing back to the Here and Now with an analysis of psychedelic influence in Electronic music (Chapter 17), the book takes a turn towards the mystical and explores the role of psychedelics in contemporary spirituality, death, and artistic expression (Chapters 18-22). The final chapter is a particularly exciting one; the story of the discovery of LSD has become legend in the psychedelic community, but the discovery of DMT has not received the same attention. Here we present the only published account of the world's very first pure DMT trip.

Every psychedelic drug that we know of acts by mimicking one or more of the body's natural neurotransmitters – chemicals that transfer or block impulses between nerves. If there is an exception to this rule, it may be DMT, since it may yet prove to *be* one of our natural neurotransmitters and thus not guilty of mimicry at all. From these small interactions between cells of the nervous system, significant changes in psychology, behaviour and brain connectivity may unfold. Without psychedelic neurotransmissions, this book, and much else besides, would not exist. Furthermore, if you will allow me to speak poetically, this book itself is a compendium of neurotransmissions with each author communicating his or her thoughts to the reader through its pages. As above, so below.

Let me tell you a strange thing about the psychedelic community: one does not really know what, precisely, one's target audience is. I hope that you find something of value in this book, whatever your background may be, whatever profession you have chosen, and whatever you think about the use of psychedelic compounds. We publish this book at an exciting time. Society has needed psychedelic consciousness for as long as it has been without it, and now, in the third epoch of psychedelic research, we are witnessing a surge of scientific and popular interest in this area. Let us hope that our society has grown mature enough to accept these compounds, these tools of learning, of accepting, of healing.

Yours,

Dave King

Lead Editor & Co-director of Breaking Convention

Hanson, D., Godtland, E. & Krassner, P. (2014). *Psychedelic Sex.* Taschen, Köln.

Hoffer, A. & Osmond, H. (1967). *The Hallucinogens.* Academic Press, New York.

Jay, M. (2013). Dreaming While Awake. *London Review of Books,* Vol. 35, No. 5, 7 March 2013. Available at: http://mikejay.net/dreaming-while-awake/

Lewin, L. (1964). *Phantastica: Narcotic and stimulating drugs, their use and abuse.* Dutton, New York.

Schultes, R. E., & Hofmann, A. (1979). *Plants of the gods: origins of hallucinogenic use.* McGraw-Hill, New York.

Sessa, B. (2012). *The psychedelic renaissance: Reassessing the role of psychedelic drugs in 21st century psychiatry and society.* Muswell Hill Press, London.

Shulgin, A., (1978). Psychotomimetic Drugs: Structure-Activity Relationships. *Handbook of Psychopharmacology,* Volume 11: Stimulants (Chapter 6). Eds. Iversen, L. L., Iversen, S. D., Snyder, S. H. Plenum Press, New York.

DMT & THE TOPOLOGY OF REALITY

ANDREW R. GALLIMORE

"If the outside world fell in ruins, one of us would be capable of building it up again, for mountain and stream, tree and leaf, root and blossom, all that is shaped by nature lies modelled in us."
Hermann Hesse

It's quite remarkable that Hermann Hesse wrote these words almost a hundred years ago, in his semi-autobiographical novel, *Demian*, and yet they beautifully capture our current understanding of the brain's role in modelling the world – all that appears in the world is modelled in our brains. It is no coincidence that the novel's protagonist, Emil Sinclair, describes growing up in what he calls a *Scheinwelt* – a world of illusion. Hesse was deeply fascinated by Eastern thought and, according to Vedic and Buddhist philosophy, the phenomenal world is indeed an illusion (*maya*). It is comforting to see the world around us as being somehow fixed, solid and, most importantly, *real*. But it only takes a lungful of N,N-dimethyltryptamine (DMT) to shatter this delusion. Whether the external world-in-itself, the noumenal world, is truly *real* is difficult to answer and, for the purposes of this discussion, it really doesn't matter.

The only world we can ever experience is the phenomenal world – the world that appears to consciousness. As far as we know, the phenomenal world is never transcendent – it never reaches out and touches the noumenal world; it is always in the head. Thomas Metzinger (2009) expresses it clearly:

> *"The global model of reality constructed by our brain is updated at such great speed and with such reliability that we generally do not experience it as a model. For us, phenomenal reality is not a simulational space constructed by our brains; in a direct and experientially untranscendable manner, it is the world we live in."*

For most people and for most of the time, this phenomenal world appears stable and predictable, but only because the brain has evolved to generate a stable and predictable model of the noumenal world. However, psychedelic drugs, such as DMT, LSD and psilocybin, among others, not only show us that the phenomenal world can become fluid, unpredictable and novel, but that it can be annihilated in an instant and replaced with a world altogether stranger than anything we can imagine. It is tempting to regard such perceptual aberrations as just that – 'tricks of the mind', hallucinations, illusions or, if we want to appear especially smart, 'false perceptions'. But such a self-assured attitude is hard to justify, as deciding what is true and what is false about our perceptions is far from trivial. To regard the phenomenal world as a stable and fixed entity is really just an approximation and as we begin to discover and explore worlds of astonishing beauty, complexity and strangeness, this approximation becomes less and less useful as a general model of our reality. Whilst the consensus model of reality is certainly the most informative from an adaptive standpoint, there is no reason to assume that it is the only informative model in an absolute sense and so no reason to dismiss those versions of reality that transcend our standard frame of reference.

The phenomenal world appears as a single unified experience that cannot be broken down into its constituent parts and yet, at the same time, contains a massive amount of information that enables us to

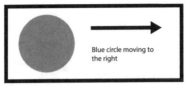

Perceived visual object

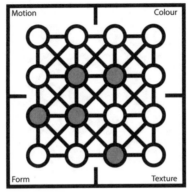

Cortical representation

distinguish each moment from the one that preceded it. Every single moment of our lives, whether waking or dreaming or under the influence of a psychedelic drug, is different from the last. This might seem obvious, but is only the case because the brain is capable of generating a practically infinite number of conscious moments, worlds if you like, each different from the last.

The brain achieves this using the principle of *functional segregation*, which refers to the way specific areas of the cortex are responsible for specific functions (Nelson et al., 2010). As we are primarily visual creatures, we will use an example from the visual system to explain how this works. To generate a visual scene, different areas of the cortex have specific roles in representing different features of the phenomenal world. There are specialised regions devoted to orientation, direction of motion,

FIGURE 1: Functional segregation in the visual cortices

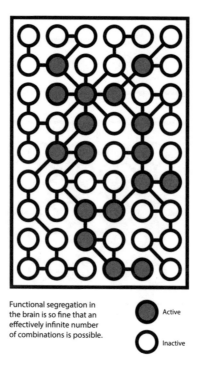

Functional segregation in
the brain is so fine that an
effectively infinite number
of combinations is possible.

● Active

○ Inactive

colour and form, for example. The primary visual cortex (denoted V1) sits at the back of the brain, in the occipital lobe. It is this region that receives visual information from the external world first, from the retina at the back of the eye and via the thalamus. V1 is generally responsible for basic visual features, containing 'simple' neurons that are tuned to respond to specific line orientations and spatial frequencies, as well as more 'complex' neurons that only respond when a line is moving in a specific direction, for example (Snowden, Thompson, & Troscianko, 2006). From V1, the information is passed to the visual association cortex, which contains areas specialised to represent specific features of the world, such as geometric shapes, colours and spatial depth. Further downstream are

FIGURE 2: Pattern of activation of functionally segregated areas of the cortex

areas specialised for the recognition of specific types of objects, such as faces or animals.

As a simple example, we can imagine how a highly simplified brain, containing just a few functional areas, would represent a simple object, such as a blue circle moving from left to right. The outline and shape of the circle are represented in one area of the cortex, whereas its colour and movement are represented in the areas specialised for those specific functions. In a real human brain, the mechanism is analogous – different functionally segregated areas of the cortex are responsible for mapping the basic features, such as the edges, contours and their orientations, as well as the overall form of the object, its colour and movement (FIGURE 1). All of these individual characteristics, each represented by a specific functional region of the cortex, when combined, define a moving blue circle, which is itself a pattern of activation in the cortex. Complex objects can be represented by specific patterns of activation of functionally specific areas of the cortex and, overall, the world that appears in consciousness is an extraordinarily complex pattern of activation of functionally segregated areas of the cortex.

The functional segregation is, of course, much finer than the simple gross distinctions between shape, colour, movement, etc. In fact, the human cortex is often described as a mosaic of minimal functional areas known as thalamocortical columns (the thalamo- prefix is from *thalamus*, the hub structure in the centre of the brain and with which the cortex is heavily interconnected). These are thought to be the basic unit of functional segregation in the brain (Hirsch & Martinez, 2006). It is the patterns of activation of these columns, or 'thalamocortical states', that represent the informational structure of the phenomenal world from moment to moment – the world is built anew with every moment, each thalamocortical state different from the last. The cortex comprises billions of columns, allowing it to generate a practically infinite number of phenomenal worlds (FIGURE 2). However, it is obvious that the brain tends to adopt a very specific set of thalamocortical states; these are the states that model the consensus world. In fact, even when dreaming, the brain tends to model the consensus world as a default. The majority of dreams are of this world and most dream activities are those from waking life (Schredl & Hofmann, 2003). The thalamocortical states that generate the dream state are identical to those of waking – the brain is modelling the world in exactly the same way, despite having no access to sensory data from the external world. This raises the question as to why the brain is capable of doing this with such adroitness. The answer is rather simple – even when awake, only a small fraction of the information used to model the world actually comes from sensory data, known as *extrinsic information*. Most of the information results from ongoing activity of the thalamocortical system, known as *intrinsic information*. Sensory data doesn't create the world, but rather, modulates this ongoing activity by being *matched* to it (Tononi, Sporns & Edelman, 1996; Edelman, 2000; Sporns 2011). The waking phenomenal world is most aptly described as a waking dream that is modulated or constrained by extrinsic sensory data (Behrendt, 2003). This is perhaps a little surprising, but explains why the brain can model the world with such expert precision during dreaming. The brain's model of the world is generated by intrinsic thalamocortical activity, modulated but not created by sensory data (Llinás, Ribary, Contreras, & Pedroarena, 1998). To understand how the brain evolved to model the world so effortlessly, we need to understand a little more about the structure of the thalamocortical system and its neuroevolution.

So far, we have considered the thalamocortical columns of the brain as independent structures. However, in reality, the columns are heavily interconnected and the activation of a single column can influence the activity of many columns to which it is either directly or indirectly connected. This connectivity is a key feature of the thalamocortical system, because it allows the brain to generate specific patterns of activation (intrinsic activity) that generate coherent and meaningful models of the world. As discussed earlier, a practically infinite number of activation patterns of the thalamocortical system are possible, but only a subset of these would represent meaningful and informative percepts. By controlling the connectivity between the columns, the brain can control the activation patterns that are generated and thus render the model of the world stable, predictable and informative.

Without this well-defined connectivity, the activity would become uncontrolled and disordered (a high *entropy* state – note this for later). If we assume the brain is a product of evolution, we can explain the development of thalamocortical connectivity using Edelman's 'Neural Darwinism' (Edelman, 1993). As patterns of sensory data are sampled from the environment, they activate specific column populations and the connections between them are strengthened, whilst others may be weakened. Over time, those patterns of connectivity that are most adaptive to survival (i.e. that generate useful models of the world) are selected and the brain gradually develops the ability to generate a stable, predictable and, most importantly, adaptive model of the world (FIGURE 3). It is a common misconception that the best phenomenal model of the world is the 'truest' one (i.e. the one that most closely matches the external world). However, the brain has no interest in truth per se and the best model is simply the most adaptive one. Mark, Marion and Hoffman (2010) used standard tools of evolutionary game theory in a simulated world to test whether *true* perceptions are always the most adaptive. The results show that this is not the case:

> "*Natural selection can send perfectly, or partially, true perceptions to extinction when they compete with perceptions that use niche-specific interfaces which hide the truth in order to better represent utility. Fitness and access to truth are logically distinct properties. More truthful perceptions do not entail greater fitness.*"

This is an important result, meaning the phenomenal world is a functional model and not necessarily the truest model of the world. It is thus presumptuous to call hallucinations 'false perceptions'. A better description would be 'non-adaptive perceptions', which may or may not be true representations of the world (Hoffman, 2011).

This model of neuroevolution can also be explained using information theory. If we imagine an early (and purely hypothetical) brain that hasn't yet evolved to model the external world, this brain will still be spontaneously active and this activity will contain intrinsic information (and perhaps even generate a phenomenal world). This intrinsic self-

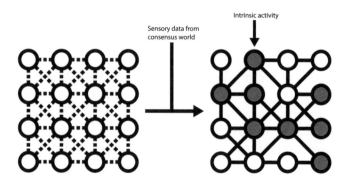

information can be quantified as the *entropy* of the brain. However, this information tells the brain *nothing* about the environment (external world), which itself contains a quantifiable amount of information (entropy). As the brain evolves by sampling sensory information from the environment and the connectivity of the thalamocortical system is moulded, the spontaneous activity of the brain becomes more and more informative (from an adaptive perspective) about the external world. In information theory this is known as an increase in the *mutual information* between the brain and the environment. The mutual information between two variables (in this case the brain and the environment) is a measure of how much we can learn about one variable by knowing something about the other. This can be represented using a set diagram (FIGURE 4); the entire entropy (information content) of the brain is one circle, that of the environment the other. The overlap between the circles is the mutual information between them. As FIGURE 4 shows, the overlap gradually increases with the progression of neuroevolution, as the brain develops the ability to generate an informative and adaptive model of the world. The term entropy is often associated with disorder, but it is actually a measure of the number of possible states of a system.

FIGURE 3: As sensory data is sampled from the environment, connections between cortical columns are strengthened and weakened. Those connectivity patterns that generate the most adaptive model of the external world are selected. Eventually, the intrinsic activity of the thalamocortical system models the consensus world as a default. (Only strong/characteristic connections are shown.)

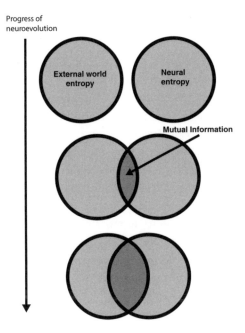

With regards to the brain, we can think of this as the number of possible thalamocortical activation patterns or states. As we have seen, the number of states, whilst theoretically almost infinite, is limited by the connectivity of the system. The brain evolves to *minimise* the overall entropy of the brain, such that the world remains stable and predictable (Friston, 2010), whilst concomitantly *maximising* the mutual information between itself and the environment.

We are now in a better position to bring all of these ideas together and formulate a generalised model of phenomenal reality and its relationship to the action of psychedelic drugs, including DMT. We can begin with the idea that the brain is capable of generating a limitless number of thalamocortical activation patterns or thalamocortical states. If all

FIGURE 4: As the evolution of the brain's ability to build an adaptive model of the environment progresses, the mutual information between them gradually increases

possible states are plotted as points on a 2D plane, the human brain's entire phenomenal reality potential exists on this plane, with each point on the plane representing a single thalamocortical state (FIGURE 5). Assuming all phenomenal worlds that can be experienced must have an informational representation in the thalamocortical system, it is not possible for a human to experience a reality that is not represented on this plane. As an individual can only occupy a single thalamocortical state at any moment, each point on the plane actually represents an individual's entire world, or *reality tunnel*, at that particular moment. An individual moves through time by shifting between thalamocortical states, represented by moving around the plane. Whilst the plane contains an effectively infinite number of points, it is not flat, as all thalamocortical states are not equally accessible. As we have seen, consensus reality is the set of thalamocortical states that have been selected during evolution by the moulding of its connectivity. These accessible states are low-energy states that can be represented by wells and valleys in the reality plane. High-energy states are those that, under normal conditions in healthy individuals, are inaccessible to the thalamocortical system. This reality plane thus begins to resemble an *attractor landscape* (Goekoop & Looijestijn, 2011) – the thalamocortical system is pulled towards the low-energy states that generate a model of the world that is most adaptive, but this set of states must not be regarded as representing the real world-in-itself. It is merely a subset that happens to possess properties adaptive to the survival of the individual and our species (i.e. represent the most informative and adaptive model of the world). The topology of this attractor landscape is determined by the thalamocortical connectivity of the individual brain and can be regarded as the topology of an individual's reality. This topology will naturally vary between individuals. In most people, the adaptive states are low-energy states and non-adaptive states are inaccessible. However, if the development of an individual's connectivity, and thus the topology of their reality, progresses abnormally, it is easy to see how non-adaptive states may become low-energy attractors. The thalamocortical system would then be able to access these normally energetically forbidden

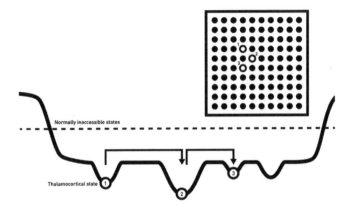

states and this would be experienced as hallucinations and other deviations from consensus reality. Schizophrenia is now described as a 'disorder of connectivity' (Tononi & Edelman, 2000) and such reality-distorting conditions may be regarded as a maladaptive reality topology.

Psychedelic drugs, including LSD, psilocybin and DMT, appear to dramatically increase the number of states available to the thalamocortical system. Although the details of how this is achieved are still being studied (Nichols, 2004; Carhart-Harris et al., 2012), these drugs bind to a specific subtype of serotonin receptor, known as the 5-HT_{2A} receptor. Whilst the brain's connectivity might be compared to the wiring of an electronic device, it is actually extremely plastic and finely balanced – this enables the connectivity to change throughout life on timescales from minutes to years. By selectively activating the 5-HT_{2A} receptor, psychedelic drugs appear to alter the reality topology itself, such that the energetic barriers to normally forbidden and novel states are lowered

FIGURE 5: All potential states of the thalamocortical system can be mapped onto a 2D attractor landscape (right). A single state is occupied at each moment in time and represents the entire phenomenal reality to an individual. From the side (below) it is clear that not all states are equally accessible and the system tends to adopt low-energy attractor states determined by the connectivity of the thalamocortical system. Consequently, most individuals, unless under the influence of a psychedelic drug or related substance/practice/disease that manipulates the topology of the attractor landscape, will only ever experience a small subset of the potential states and thus a highly limited version of reality, known as the consensus world

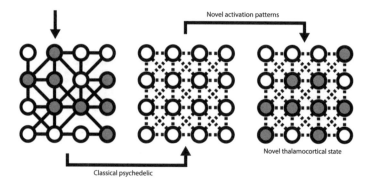

and can be explored (FIGURE 6). Carhart-Harris et al., (2014) propose that these drugs increase the entropy of the brain, which is equivalent to stating that the number of accessible thalamocortical states is increased, as suggested here. Interestingly, this is an effect that ought to be measurable using modern functional neuroimaging techniques. As the states accessed by the brain during a psychedelic experience are no longer limited to the most adaptive states, there will be a decrease in the mutual information between the brain and the environment. This is illustrated in FIGURE 7 and it is notable that the process appears to be the reverse of the progress of neuroevolution. Carhart-Harris et al., (2014) suggest that the psychedelic experience is a primitive state of consciousness that preceded the development of normal waking consciousness seen in modern humans and this makes sense from this perspective.

DMT is exceptional in that, given a sufficient dose, the separation of brain and environment becomes complete and their mutual information is reduced to zero. The individual's phenomenal reality is completely replaced by an entirely alien reality that is unrelated to the consensus world. In fact, the brain actually loses its ability to sample data from the consensus world and render a meaningful percept – the user will normally -lie back, close his/her eyes and hold tight. It remains a matter

FIGURE 6: By activating 5-HT$_{2A}$ serotonin receptors, classical psychedelics facilitate the adoption of novel activation patterns by the thalamocortical system

of debate as to whether this bizarre world is an autonomous and objective reality (for a full discussion, see Gallimore, 2013), but it is difficult to explain why the brain is capable of suddenly generating phenomenal worlds of such beauty and complexity that bear no relationship whatsoever to the consensus world, which, as far as we know, is the only type of world the brain has evolved to model.

Of those that manage to inhale sufficient DMT to 'break through', a significant proportion describe eerily similar types of experience – hypertechnological alien-like worlds, complex machinery, insectoid and mischievous elf-like creatures and other motifs characteristic of the DMT experience (Strassman 2001; Luke 2011). This has convinced many that the DMT reality must have an objective existence. However, such a conclusion raises more questions than it answers. If the DMT world *is* real, how is the brain able to receive, parse and render data from such a reality? Unless the DMT reality is modelled by the brain in an entirely different way to consensus reality, which there is absolutely no evidence for and seems completely unparsimonious, we must assume that it has an informational representation in the brain and is constructed using intrinsic activity of the thalamocortical system. The question is whether this activity is modulated by an extrinsic component, as is the consensus world. It is evident that the drug alters the thalamocortical system's intrinsic activity such that the mutual information between the brain and the consensus world is reduced to zero. Does the mutual information between the brain and the DMT reality concomitantly become non-zero (i.e. is there a sensory interaction between them)? There seems no obvious mechanism for the brain to interact with an alternate reality and we must leave this as an open question. However, the DMT experience is a stark demonstration of the brain's ability, under specific conditions, to explore regions of the thalamocortical reality topology that, without DMT, we might not suspect even exist.

As humans, we are equipped with a brain capable of modelling and experiencing a vast multitude of realities. This is not a hypothesis to be tested; it is a fact demonstrable to anyone willing to inhale two or three lungfuls of DMT. Consensus reality is very much a functional reality,

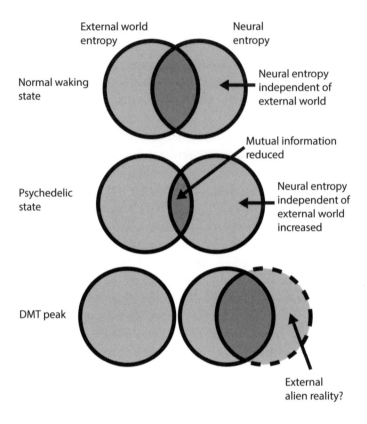

a phenomenal reality model that the brain has evolved to facilitate its survival in the noumenal world. It is all too easy to assume that consensus reality is a privileged model of reality, the truest model, *the real thing*. But, as we have seen, it is actually just a tiny subspace within a much broader reality topology available to the brain's extraordinary information-generating machinery and, with the aid of a select number

FIGURE 7: By increasing the number of available thalamocortical states, psychedelic drugs increase the entropy of the brain independent of the external world, whilst decreasing the mutual information between them. With DMT, the mutual information (between brain and environment) is reduced to zero. It is unknown whether there is a concomitant increase in the mutual information between the brain and an objective alternate reality

of natural and synthetic psychedelic drugs, this entire topology may become accessible and open to exploration.

Behrendt, R. P. (2003). Hallucinations: Synchronization of thalamocortical γ-oscillations underconstrained by sensory input. *Consciousness and Cognition*, 12, 413–451.

Carhart-Harris, R. L., Erritzoe, D., Williams, T., Stone, J. M., Reed, L. J., Colasanti, A., Tyacke, R. J., Leech, R., Malizia, A. L., Murphy, K., Hobden, P., Evans, J., Feilding, A., Wise, R. G. & Nutt, D. J. (2012). Neural correlates of the psychedelic state as determined by fMRI studies with psilocybin. *Proceedings of the National Academy of Sciences USA*, 109(6), 2138–2143.

Carhart-Harris, R.L., Leech, R., Tagliazucchi, E., Hellyer, P.J., Chialvo, D.R., Feilding, A., Nutt, D. (2014). The entropic brain: A theory of conscious states informed by neuroimaging research with psychedelic drugs. *Frontiers in Neuroscience*, in press.

Edelman, G. M. (1993). Neural Darwinism: Selection and re-entrant signalling in higher brain function. *Neuron*, 10, 115–125.

Edelman, G. M. (2000). *A Universe of Consciousness: How Matter Becomes Imagination*. Basic Books.

Friston, K. (2010). The free-energy principle: a unified brain theory? *Nature Reviews Neuroscience*, 11, 127-38.

Gallimore, A. R. (2013). Building Alien Worlds - The Neuropsychological and Evolutionary Implications of the Astonishing Psychoactive Effects of N,N-Dimethyltryptamine (DMT). *Journal of Scientific Exploration*, 27(3), 455-503.

Goekoop, R. & Looijestijn, J. (2011). A Network Model of Hallucinations. *Hallucinations: Research and Practice* by Blom, J. D. D. & Sommer, I. E. C. New York: Springer.

Hirsch, J. A, & Martinez, L. M. (2006). Laminar processing in the visual cortical column. *Current Opinion in Neurobiology*, 16(4), 377–384.

Hoffman, D.D. (2011). The Construction of Visual Reality. In *Hallucinations: Research and Practice* by Blom, J. D. D. & Sommer, I. E. C. New York: Springer.

Llinás, R., Ribary, U., Contreras, D. & Pedroarena, C. (1998). The neuronal basis for consciousness. *Philosophical Transactions of the Royal Society of London B*, 353, 1841–1849.

Luke, D. (2011). Discarnate entities and dimethyltryptamine (DMT): Psychopharmacology, phenomenology and ontology. *Journal of the Society for Psychical Research*, 75, 26-42.

Mark, J. T., Marion, B. B. & Hoffman, D. D. (2010). Natural selection and veridical perceptions. *Journal of Theoretical Biology*, 266, 504-515.

Metzinger, T. (2009). *The Ego Tunnel*. Basic Books.

Nelson, S. M., Cohen, A. L., Power, J. D., Wig, J. D. P., Miezin, F. M., Wheeler, M. E., Velanova, K., Donaldson, D. I., Phillips, J. S., Schlaggar, B. L. & Peterson, S. E. (2010). A parcellation scheme for the left lateral parietal cortex. *Neuron*, 67, 156–170.

Nichols, D. E. (2004). Hallucinogens. *Pharmacology and Therapeutics*, 101, 131–181.

Schredl, R. & Hofmann, F. (2003). Continuity between waking activities and dream activities. *Consciousness and Cognition*, 12, 298–308.

Snowden, R., Thompson, P. & Troscianko, T. (2006). *Basic Vision: An Introduction to Visual Perception.* Oxford University Press.

Sporns, O. (2011). *Networks of the Brain.* MIT Press.

Strassman, R. J. (2001). *DMT: The Spirit Molecule.* Park Street Press.

Tononi, G., Sporns, O. & Edelman, G. M. (1996). A complexity measure for selective matching of signals by the brain. *Proceedings of the National Academy of Sciences USA*, 93, 3422–3427.

Tononi, G., Edelman, G. M. (2000). Schizophrenia and the mechanisms of conscious integration, *Brain Research Reviews*, 31, 391-400.

GILLES DELEUZE AND PSYCHEDELIC THOUGHT AS RESISTANCE

OLI GENN-BASH

The philosophy of Gilles Deleuze attempts to challenge preconceptions regarding thought, our experiences, and reality. This challenge manifests as resistance that occurs in the mind, through a revolution in the way in which we think. This resistance occurs in response to what Deleuze perceives to be a reduction of thought itself to externally imposed orders within traditional philosophical thought. This implies the creation of dominant narratives, which create presuppositions resulting in a limitation to the commencement of thought. For Deleuze, these presuppositions must be dispelled in order for thinking to actually happen; as the problematic nature of these presuppositions is that they are subjective and exist for so long without serious evaluation. Deleuze's conscious creation of *The Image of Thought* as a concept provides an extensive and concise evaluation of what is seen to be the dogmatic nature of traditional philosophical thought, which must be resisted. I will explore resistance to this concept within the context of psychedelic experiences, particularly within the context of my personal experiences with LSD – I will later explain why I have chosen to focus on this specific substance, but first I will further explore the philosophical foundation of the *Image of Thought*.

DELEUZE AND THE IMAGE OF THOUGHT

Deleuze challenges the presupposition that traditional philosophical thought contains no presuppositions. The example of Descartes 'Cogito' is initially used to clarify this point further: There is an attempt made by Descartes to escape the definition of the rational nature of humans, in order to evade all the objective presuppositions which restrict those processes which function by means of classification. According to Deleuze however, the presuppositions that are imbedded or subjective cannot be avoided, as they exist not within concepts but rather in opinion. Ultimately there is the belief that everybody knows what is meant when we discuss the 'self', 'thinking', or 'being', irrespective of any concepts regarding these notions. The idea of 'I think' as foundation of the fundamental self only appears in such a way because the presuppositions have been referred back to the experiential self. To uncover what is meant by the imbedded or subjective presuppositions, Deleuze focuses on this idea of 'everybody knows' outside of a conceptual or philosophical context. When Descartes says 'I think therefore I am', there is the assumption that everybody will understand what thinking and being means, which no one can deny (Deleuze and Patton, 2005).

This presupposition ultimately allows for a dogmatic, rather than inclusive, process of philosophical thought which claims however to commence without any presuppositions whatsoever. Deleuze's problem with this notion is an extreme moment of resistance, seemingly shaking one out of this comfort. Arguably it is resistance to what has been laid out as a comfortable and secure precedent over a proceeding statement, allowing the thought to be cemented and then forcing the individual to become represented in the wider scope of 'everybody'.

To be an individual without presuppositions requires an active attempt to avoid representing anything, or being represented. In possessing the humility of not being able to know what 'everybody knows', along with the denial of what everybody is supposed to identify, one is able to enact a fervent form of resistance against a dogmatic presupposition. Philosophical thought puts forward what is meant by thinking, being, and self to be given universal recognition. The focus is therefore on the shape of representation or recognition in general. For Deleuze, this form of representation or recognition contains the presupposition that we are naturally capable of thought, with an innate capacity always inclining towards truth or the true, as the philosopher themselves are endowed with good will and utilise the supposed upright nature of thought. However, this point is hidden because the proposition is that within conceptual philosophical thought there is a presupposition of a pre-philosophical 'Image' of thought, taken from constituent of pure common sense. Rather than putting forward various images that correspond to the philosophy in question, it is a single Image in general which Deleuze sees as that which composes the "subjective presupposition of philosophy as a whole." What he is suggesting is that for philosophy to escape the presupposition of being supported by an Image that inclines towards the truth, it must uphold an essential critique of this Image as its foundation. In resistance to this Image, philosophical thought must continue within thought outside of it, abandoning the form of representation and the notion that 'everybody knows', whilst accepting the paradoxical nature of things. This liberation from the image allows for the commencement of thought to be thought, with a constant repetition of this commencement when freed from the imposed directives of traditional philosophical thought (Deleuze and Patton, 2005).

The end of this paragraph is particularly interesting when looking at what one could view as the mystical quality to Deleuze's conception of the Image, and how philosophical thought can be liberated from it through certain practices. It is not so much that he provides a substantial remedy to the problem that he has set out. Rather, there is an identification of this Image-less thought against an understanding of what it is not. One may be able here to draw comparisons between this description and ideas within Zen Buddhist thought, particularly when looking at the concept of the *Koan* and the idea of resistance to recognition.

The Zen *Koan* is a riddle that seeks to shake one's mind out of its ordinary thought patterns, creating non-rational connections between the descriptive aspects of the riddle and that to which the linguistic expression refers. It is an attempt to produce a different state of mind and 'baffle the intellect', rather than to propose a logical problem that must be solved through an employment of one's own understanding or intelligence. There is the story of a Zen master addressing a monk on the topic of a bamboo stick: "The Zen master asks, 'If you call this a stick, you affirm; if you call it not a stick, you negate. Beyond affirmation and negation what would you call it?'" (Zug III, 1967). Much like Deleuze's notion of a thought without Image, the Zen *Koan* can be viewed as that which causes one to become comfortable with and internalise paradoxes, whilst dispelling the presupposition that there is something to be figured out or truth to be found in what has been proposed. However, there must be further exploration into the imposed directives, which construct the Image, and the way in which Deleuze sees resistance to this Image through his understanding of thought.

As mentioned before, there is an attempt to dismiss the notion of thought as an innate capacity, with thinking being done in an upright and good-natured fashion, which inclines towards the uncovering of truth. Rather this thinking happens, for the most part, as a result of an impulsive response to something. This notion of thought being involuntary is crucial to understanding how the Image can, and should, be resisted. This involuntary thought is a production of what Deleuze views as the *encounter*. This *encounter* is seen to be that which can only be sensed, and is placed in opposition to the notion of thought as recognition. The notion of recognition utilises the faculties of experience as a means to continue the construction of the common sense. (Deleuze and Patton, 2005).

PSYCHEDELIC THOUGHT AS RESISTANCE TO THE IMAGE OF THOUGHT

In his critical guide and introduction to *Difference and Repetition*, James Williams looks at Deleuze's concept of the 'Idea' as a result of the encounter. The Idea is different from the notion of *ideas*, meaning any identifiable thought about something. The Idea is what must be connected to through the encounter, as it provides the Image-less condition for thought, which animates concrete lives through intensity and difference. What is encountered may be sensed in various ways, but it can only be encountered by sensing. This concept of the Idea is a way to explain the relationship between things external to the mind, which provides the space for our ideas to evolve and change. This relationship is in constant flux, with no fixed character, and can only be sensed and known to a degree through the encounter. What is being resisted is the need to preserve identification. The Idea is comprised of difference and intensities that can only be inferred from the notion of sensation as a basis (Williams, 2003).

This notion of the Idea as a relationship that is constantly changing without any permanent identity seems to suggest an almost meditative method of uninhibited interaction, which allows for the encounter to happen. Although Deleuze does not point towards any drug-induced or mystical experiences, I view this uninhibited interaction as being representative of a psychedelic experience, particularly one that has been induced through use of LSD. There is a difficulty in analysing the psychedelic experience in general as a 'psychedelic experience' might contain many different aspects, or mean various things to different people. In this instance I do not necessarily see any merit in exploring other subjective viewpoints regarding the psychedelic experience, as there is the very likely possibility of just falling into the trap of a psychedelic *image* that others have created. This would not really allow for an exploration into psychedelic thought as resistance to the *Image of Thought*, as we would merely be viewing this resistance in a confined manner, which is exactly the opposite of what Deleuze is putting forward. I will therefore be exploring my personal experiences with LSD, and comparing my conception of thought within this state of mind to Deleuze's conception of resistance to the *Image of Thought*.

I do not see the LSD experience as something that necessarily 'provides' anything on top of our ordinary state of consciousness – however I do view this particular experience as having an immense impact on thought nonetheless. In fact, it may be the case that this experience is doing the opposite, and actually taking away from our normal everyday consciousness. I put forward the point that through engaging in this experience the usual confines of thought itself are removed. As previously mentioned, these confines are within the context of recognition and ownership. For example, one recognises the object which they tend to place items on as a 'table', and the thought of this table is then individualised and owned as that person 'knows' what a table is. We can therefore go about our daily lives comfortably, recognising and owning within the context of thought. However, when one encounters a table within the midst of an LSD experience a completely different process might occur. Rather than putting something down on the table to utilise the table-like functionality, an individual under the influence of LSD may be solely concerned with exploring the patterns in the grains of the wood. One could be exploring the shape of the table and the patterns for hours before even noticing that it is a 'table'. We may feel as if we have interacted with the world for the very first time, almost entering back into a childlike state where we are captivated by every little aspect of our reality. The removal of filters and preconceptions that have been built up during our adult lives allows for thought to occur without any restrictions. This may become problematic if we our too concerned with holding onto these constraints for fear of being uncomfortable. I believe that something such as the LSD experience provides an immense opportunity for thought to resist the propensity towards dogmatism, in both a philosophical and general sense.

However, whilst the psychedelic experience can allow one to become more open, one must become open to the rejection of thought manifesting as recognition or identification in order for the *Image* to be resisted. The indication of an uninhibited method of interaction however, seems to require an openness as a prerequisite to having this encounter. Ultimately we are faced here with a paradox that must be grappled with. How can one simultaneously be open to an encounter whilst requiring the encounter to produce the openness? Is Deleuze assuming that we are already open to such an encounter? The extent to which we can make ourselves open to this encounter becomes very complicated. If we are indeed to *make* ourselves open to this, then we are closing down the possibility of the uninhibited interaction by willing it. There is the possibility however, that by chance we are open to this interaction.

This Deleuzian understanding of resistance to the *Image* ultimately requires a combination of passivity and activity. It is the notion of attempting to makes one's self as open as possible. But already we have raised the problem regarding an attempt to *make* the self as open as possible, as it could contradict the very concept of openness through constructing the means by which to be open. So rather than an opening, it is a construction which closes down or distils the very nature of what it means to be 'open'. However, this critique can merely be viewed as that which falls into the presupposition of what it means to think. By raising a problem with the paradoxical nature of this conception we are allowing our thought to become dogmatised by an *Image,* which seeks to find concrete logical answers and truth. For thought to escape dogmatism it must occur, more often than not, when there is no active attempt to think – within a passive construction of the self.

CONCLUSION

How we come to understand what it means to think requires a transformation of ordinary thought patterns. In order to even begin to think about what it means to think, we must acknowledge that what is perceived to be common sense must be challenged.

It is as if we must utilise the moments where we are passive as tools that enable this openness to the encounter. Be it within different aspects of our lives, which may seem abstract or arbitrary to our perception of thought. However, there is a quality to these aspects that we are not aware of until we step back and attempt to understand the nature of thought. We may view thought itself as an art form, not needing to control or impose, but simply as a means to engage with the world. It is a means for the world to reveal itself. Engaging in psychedelic experiences such as that induced by LSD provides the opportunity for thought which has the potential to lie outside the realm of our normal thought patterns of what is immediately tangible. We may see different and abstract patterns in art, hear new sounds within a piece of music, discover new feelings in a relationship, or have our visual and sensory perception changed within such an experience. Or it may be as simple as the time we spend in the shower every morning completely sober. When we stop trying so hard to think, almost naturally, we may find that is when our mind is calm enough and uninhibited to the point where an Image-less moment of thought through interaction may occur. Some may constantly desire to explain the inexplicable, they may be too attached to categorisation and ownership over understanding. One may constantly question what the point of all this is, and claim that Deleuze has merely led us down a strange metaphysical path only to create his own Image. This rigidity is ultimately undesirable, and should be resisted at all costs! It fails to grasp or acknowledge the potential of thought to be thought about in a different way.

The interaction does something to our mind that causes us to think differently. We should actively seek to engage with that which allows for the passivity, creating the openness required. This is a paradox, which has been grappled with – the simultaneous existence of passivity and activity, with one essentially feeding into the other. We utilise the activity to allow for the passivity, but the passivity opens us up for the activity. Arguably the acceptance of this paradox is a moment of thought resisting its own tendency towards dogmatism. We must relinquish our perceived notion that we understand everything in a straightforward manner. The problem of the Image proposing concepts regarding the way in which the world ought to be becomes problematic if the world takes us by surprise. The *Image* places humans above everything, separate from nature, as if we are almost guardians of the divine knowledge of the universe. In resisting this *Image* however, we allow ourselves to be in the world, interacting as part of this changing flow of encounters.

Deleuze, G., Patton, P. (2005). *The Image of Thought* in *Difference and Repetition* – Continuum International Publishing Group.

Williams, J. (2003). *Gilles Deleuze's 'Difference and Repetition': A Critical Introduction and Guide* – Edinburgh University Press

Zug III, C. G. (1967). The Nonrational Riddle: The Zen Koan – *The Journal of American Folklore,* Vol.80, No.315 – American Folklore Society

INTERPRETING PSYCHEDELIC CONSCIOUSNESS THROUGH BERGSON'S PROCESS PHILOSOPHY

PETER SJÖSTEDT-HUGHES

The main points I wish to convey in this concise text:
- That all existence is movement: 'Process Philosophy'
- That language is extremely useful because it is extremely misleading
- That the understanding has evolved for practicality (power) rather than for knowledge
- That brain does not produce the mind, despite neural correlates of consciousness
- That the observer *is* the observed (that subject and object are partially one)
- That mind is ubiquitous: 'Panpsychism'
- That fundamental reality is metaphysical creativity – *and* that this can be intuited through psychedelic intake

In effect, I shall endeavour to show that the re-emerging 'Process

Philosophy' of Henri Bergson et al. offers a potent lens through which the psychedelic state can be understood.

PROCESS PHILOSOPHY

Process Philosophy is a term notably addressed retrospectively to the Nobel laureate philosopher Henri Bergson (1859-1941), and to the Cambridge philosopher-mathematician, A. N. Whitehead (1861-1947) who explicitly borrowed some of his ideas from Bergson. Bergson emphasised the fact that if we consider any so-called object we realise that it has no definite boundaries *in space or in time*. For instance, a 'mushroom' constantly changes its form, and is in fact one with its mycelia (roots), which in turn is one with the nutrients 'it' gathers in the substratum, etc. There exists no clear boundary. Even with more durable objects, if we accelerated time and saw them over eras, we would see mountains fluctuate like waves in the ocean. Contrariwise, so-called atoms and molecules also fluctuate, of course – here modern quantum physics shadow Bergson's premeditations.

That language is extremely useful because it is extremely misleading and that the understanding has evolved for practicality (power) rather than for knowledge.

It is language, or more specifically words, which cut out of the fluidity that is reality, isolated 'things'. This isolating process (differentiation) serves the human species very well: in order to predict the future and to create tools/technology (for our survival and development), we need to assume that there are stable 'things' so that we can place them into a model, and apply to these things stable 'laws'/'constants', and expect from these practical results. And though this method of extraction and *hypostatisation* (solidification) yields the produce of *science* (medicine, weapons, etc.), *we must realise that this natural method is merely a model of reality, not reality itself.* As both Bergson and Friedrich Nietzsche (1844-1900) stated, through divergent lines of thought, we are all inbuilt *Platonists*: we mistake concepts for reality. We are especially prone to geometrical conceptualisation, mistaking fluid reality for the motion of stable things in a pure geometric space. Plato is himself the gaoler of his famous cave – escape from which may be effectuated through psychoactive intake, as we shall see.

In reality, things and space are *one entity*: movement/becoming/ change. It is difficult for the human species to think of movement without thinking that some *thing* must exist which moves (e.g. water in waves), but that is simply a result of our evolved natural mode of thought. We must ontologically prioritise movement over 'things that move', as 'things' *are* movement. It was the abandoning of the theory of aether as the substance underlying the movement of *electromagnetism* that led to Einstein's paradigmatic theories (via Maxwell). We even like to split time into 'things': *instants*. We thereby think of time as separate from (equally separated) 'space', 'matter', 'forces', and so on. Due to our practical geometric bent, we think of this differentiated time as instants on a spatial line. But an 'instant' as such cannot exist because it must have a beginning, and an end, and therefore a duration. This abstraction, this differentiation, is a mistake that leads to *Materialism* and *Determinism*: i.e. that reality can be *reduced* to 'things' (corpuscles) which act according to timeless laws ('constants') which can be thus theoretically predicted instant-by-instant. We have abstracted from experience elements which never existed separately in reality, and then we have tried to put those elements back together again to concoct phenomena such as consciousness as an *effect* (i.e. epiphenomenon) of these abstractions. Whereas in truth consciousness was *already* present prior to that abstraction effected throughout experience. This illusion is useful to mankind, yet it is *mistaking the model for reality*. Time, space, matter, force, etc. are all one phenomenon in experience, they are divided in analysis – but we must bear in mind that the analysis is artificial rather than natural. The purpose of philosophy, according to Bergson, is to think beyond this human condition of useful analysis.

In summary thus far, there can be no 'instants' (a mistake due to the spatialising and division of such spatialised time). There can be no 'things' (a mistake due to isolating repetitive movements away from their continuations). There can be no known 'laws/constants' of nature, due to the *Problem of Induction* (from David Hume): the natural belief that the unobserved resembles the observed – a necessarily non-provable axiom.

Therefore: We must acknowledge that our understanding has evolved for practicality rather than for pure knowledge. We must understand that illusion serves us well, and realise that other creatures will have their own modes of thought, useful for them – their realities thereby no doubt differing widely from ours (à la Thomas Nagel's question, What is it like to be a bat?).

That brain does not produce the mind, despite neural correlates of consciousness.

Due to this, our evolved practical mode of thought, we are naturally inclined to *reduce all phenomena to things-moving-in-time* (rather than the reality of pure movement). This is then Materialism, or Mechanism. So when addressed with the phenomenon of *consciousness*, it seems natural to reduce it to material things moving in the variable of time. Such a belief has a long history, stemming back at least to the Atomism of Leucippus and Democritus two and a half thousand years ago.

Today, more specifically and especially, consciousness is presumed to reduce to neurons firing impulses and molecules to one another within the brain.

This error (of confusing model for reality) is further entrenched within the modern mindset due to the additional error of confusing correlation with sufficient cause or identity. It is believed by many that because consciousness is correlated to brain activity (in brain scans and in brain damage), the brain must either sufficiently *cause* consciousness (epiphenomenalism) or be *identical* to consciousness (identity theory): often referred to through the notion of 'neural correlates of consciousness'.

But from correlation to cause or identity is a non sequitur (it does not logically follow) because, analogously, as Bergson points out, there is also a *perfect correlation* between a radio set and the programme it is playing. Change the radio's circuitry and you will change the perceiving of the programme. One could even predict/read the programme from investigating the radio's circuitry. But, of course, this perfect correlation does not imply that the radio sufficiently (totally) *causes* the programme! Neither does the perfect correlation imply that the radio *is* (identical to) the programme. In fact, in this case, the radio merely picks up and translates the programme, which has its source elsewhere. This is analogous to the ultimate Bergsonian view. That analogy seeks to show that mind is *not necessarily* brain, or caused by the brain. Another old, but recently revived argument, shows that the mind *cannot* be sufficiently understood by examining the brain: the *'Hard Problem of Consciousness'*. That is, that 'things' (molecules, pulses, etc.) *moving* according to procedures ('laws') can never yield the full understanding of any knowledge of subjectivity. For instance, dopamine levels may be correlated to the feeling of satisfaction, but no matter the complexity, satisfaction cannot be fully understood by a mere analysis of matter moving. There is a huge chasm between matter moving and qualia (experienced qualities). As Bertrand Russell stated (2007/1927):

> *"It is obvious that a man who can see knows things which a blind man cannot know; but a blind man can know the whole of physics. Thus the knowledge which other men have and he has not is not part of physics."*

These problems are only problems for Materialism. The problems emerge due to the original mistake of extracting from the flux of reality, separate (artificial) parts in terms of space, time and force. It is as if someone looked at a cake, extracted all the tones of colour from it, then painted it using those colours, believing that with a precise enough duplicate the painting would eventually have all the properties of the original cake, including its taste.

That the observer is the observed; (that subject and object are partially one).

So, if brain is not mind, what is? Bergson employs his Process Philosophy (that *all* is process, becoming, movement, change) and argues that the perception we have of 'something' is *actually a part of that something as well as a part of 'oneself'*. There is a continual uninterrupted flow. (Obviously the words 'something' and 'oneself' here are used metaphorically at this point, as ultimately are all words.)

Consider, say, looking at a star. If we trace the actual elements involved in this *process* we understand that electromagnetic waves of a certain frequency move towards us, this light then transforming through the eyes' lenses hitting the retina, transposing into an ionic pulse through the optic nerve to the occipital lobe, thereafter continuing to virtual (possible) bodily actions via the whole nervous system, etc.

The words 'star', 'eyes', 'brain', 'nervous system' seemingly present these concepts as isolated parts which may interact with each other. The reality is, Bergson argues, that these are all one system, with artificial cuts (necessary for utility). There is no absolute distinction between the eye, the brain and the nervous system. *So why isolate consciousness at an artificially-created part of the entire flow (at the brain)?*

Furthermore, why isolate the 'eye' from that which it perceives? The eye and the electromagnetic frequencies it redirects are part of *one process*. And the electromagnetic frequencies are part of the *star* from which they emanate, again there being no absolute distinction.

In other words, the observer is the observed. *Part of the star, and my perception of the star, are numerically identical (the same thing).* (All words in that sentence being mere artificial extractions from a single flowing reality.) That part of the star that is my perception of it, is the part that *evolution* has *extracted* for the practical purposes of me, the human. That is, again, that my perception of an object, and part of that object, are one. They lie in the relation *part-to-whole* rather than in the relation *representation-to-object* (the latter is commonly believed not only by Materialists, but also by Cartesian Dualists, Kantian Idealists, and indeed most of Western intellectual history).

Immediately before Aldous Huxley refers to Bergson in his well-known text, *The Doors of Perception* (2004/1954), he writes (of his mescaline experience):

"I spent several minutes – or was it several centuries? – not merely gazing at those bamboo legs, but actually being them – or rather being myself in them; or, to be still more accurate (for 'I' was not

involved in the case, nor in a certain sense were 'they') being my Not-self in the Not-self which was the chair."

I believe the passage expresses well a more direct intuition of our part-to-whole relationship with our environment, purified from common conceptual consciousness.

If one damages or alters part of the brain, part of the process is altered, so a concomitant alteration in mind would ensue; but likewise, alter the process elsewhere, outside the body (for example, cut the emanating light with a cloud) and the consciousness will also change. *If one scanned the brain during a star-gazing session, one would expect to find a perfect correlation between the neural correlates and the reported vision, according to Bergson's hypothesis that the brain does not produce consciousness but is merely a centre for the redirection of incoming signals (frequencies) to virtual (potential) bodily actions.* Again, such neuroscientific correlative data does therefore not by necessity suggest a materialist explanation. Technically, to assume that the mind is caused by the brain, due to psycho-physical correlation, is to commit the fallacy *cum hoc ergo propter hoc.* William James, the great philosopher, psychologist and physician, makes the same point (1898) when he states:

> *"In strict science, we can only write down the bare fact of [mind-brain] concomitance; and all talk about either [consciousness] production or transmission, as the mode of taking place, is pure superadded hypothesis, and metaphysical hypothesis at that... Ask for any indication of the exact process either of transmission or of production, and Science confesses her imagination to be bankrupt."*

This purpose of the brain, for Bergson, to direct incoming data to possible (virtual) bodily movements, gives us *power* over our environment. It is not to produce consciousness, but only to *streamline* it to practical considerations. The activity of the brain is the continuation of the movement of 'external objects' within us, so to be able to further that movement to our own purposes: in line with Nietzsche, this is ultimately to develop power over our environment.

Further to perception being a part of the 'object', Bergson states that all perception includes *memory*. For instance, to see the colour of an object involves contracting innumerable electromagnetic waves into one's 'present'. Therefore, *consciousness is essentially memory, as there can be no consciousness without a contraction of the past.* And as the brain does not produce consciousness per se, memory is *metaphysical.* It is well to note at this point that memory has been demonstrated in certain plants (Gagliano et al., 2014) and slime-moulds (Reid et al., 2012), thereby indicating the fact that memory is not dependent upon a brain.

For Bergson, the past always exists in its entirety; it is the brain that limits its recollection to practicalities, and brain damage that limits its reception, that is, its effectiveness for action. In dreams, Bergson writes, when the mind is not immediately concerned with its physical environment, the past is more open to revelation.

That mind is ubiquitous: 'Panpsychism'.

Thus, if the brain is not sufficiently responsible for consciousness, and it is only the inbuilt Platonism of our understanding that lets it seem that it is – but yet that consciousness *exists* (which is the most certain fact that one can have, beyond any scientific 'fact') – then an implication is that this flow, this movement, this becoming that is *pure reality,* is itself a form of consciousness (which is memory).

Panpsychism – that all is mind – is accepted (to an extent) by Bergson, Whitehead and an increasing number of Western thinkers. Bergson thinks consciousness, or subjectivity to be more precise, is ubiquitous only throughout non-instinctual *life,* whereas Whitehead argues that subjectivity is ubiquitous throughout *everything,* including stars and molecules. *Not* that they believe, say, a table per se to be conscious: Bergson confines subjectivity then to organisms, and Whitehead confines it to 'actual entities' and 'societies' (self-organising systems: the molecules that make up the table have a subjectivity in themselves, basic but there). Whitehead calls his thought the 'philosophy of organism', arguing that 'biology is the study of the larger organisms; whereas physics is the study of the smaller organisms' (2007/1925).

Although panpsychism, or panexperientialism, may seem prima

facie unbelievable to many, it must again be realised that the currently popular paradigm of Materialism has reached a cul-de-sac: to believe that mind is a product of matter, produced by the activity of artificially differentiated elements of reality, can only lead to the Hard Problem of Consciousness, Solipsism, and a profusion of mind-matter paradoxes.

That fundamental reality is metaphysical creativity, and that this can be intuited through psychedelics.

Bergson argues that underlying our human mode of experiencing reality exists a metaphysical domain of pure creativity. One method by which he argues for this is through the examination of evolution. Bergson revealed the problems inherent within Lamarckism and neo-Darwinism, involving the problems associated with mechanism, determinism and teleology, and offered an alternative theory ('creative evolution') whereby a vital principle strives through matter to expand itself to full consciousness. Mankind is at the pinnacle of this vital drive (*élan vital*), but he makes room for the emergence of the superman, not too unlike the theories of Nietzsche: a being who has supreme power over his environment as he can contract more of reality into qualitative elements (Bergson, 1998/1907).

This metaphysical creative domain is difficult to grasp through the intellect, but Bergson emphasises the fact that it is a mistake to believe that *intellect* is the only means to knowledge. He opposes this path to that of *intuition*, which though very difficult to pursue (as it opposes our human practical condition), can yield many advances. This is because, by transcending the 'cut-up' falsely differentiated world proposed by the intellect, it allows for many former problems and paradoxes (such as Zeno's) to be seen in a pure light, as it were – his own philosophy perhaps being a prime example.

Now, by subduing the mechanism of our nervous system, psychedelics seem to be a route of entering this state of *intuition*, uniting with this metaphysical creative domain by disassociating from the practical *will*. Psychedelics are the consolation to Bergson's lament with the route of human evolution: 'Consciousness, in man, is pre-eminently intellect. It might have been, it ought, so it seems, to have been also intuition... A complete and perfect humanity would be that in which these two forms of conscious activity should attain their full development' (1998/1907).

Let us elaborate on this psychoactive process. The ingestion of certain substances (e.g. psilocybin, mescaline, dimethyltryptamine, lysergic acid diethylamide) disrupts the everyday workings of our brain and body generally. As the brain does not produce consciousness, neither do these substances directly produce psychedelic experiences. These substances rather act by hindering normal physiological functioning by inundating relevantly susceptible synaptic clefts, thus neurons and so the nervous system as a whole. Due to this disruption, memory – which as mentioned cannot be physical – can no longer act effectively for the body: the connection between incoming perceptions cannot follow the usual physiological lines and so it is not properly contracted by memory into the stable objects we would normally experience. Instead, our consciousness/memory is, as it were, freed from having to act in a practical manner. As well as the resulting loss of articulate bodily motion, this results in a number of avenues of fresh phenomenology, experience. With eyes closed, our memory, now dislocated from bodily action, retreats back into the pure mental or metaphysical realm which consists of memories of experiences and that domain of pure creativity which, for Bergson, guides art, evolution, paradigm shifts, creative thinking generally and in fact the free advance of the cosmos itself. The psychedelic experience accords with this as trips can be both personal and radically novel: impersonal and shocking in their ineffable sublimity. Theoretically, the higher the dose, the more the physiological disruption, and so the further from the personal will the experience become, as memory will be further dislocated from bodily action to which personal memories pertain.

With eyes open in this state, an abundance of unusual phenomena take place. The continuation of objects within our physiology is either halted or haphazardly passed through where possible, dependent again on dose (high to low, respectively). A table may seem to fluctuate its shape and size, for instance. Furthermore, because memory is dislocated, such an object may not be recognised as such, as 'a table', and accordingly it may not even be registered as a distinct 'thing'. Here we can understand that psychedelics can return us to a pre-analytic comprehension that excludes artificial separations ('table'), i.e. *intuition*. Further still, the artificial separation between subject and object can be overridden and a sense of unity with the so-called objects can be intuited, as quoted from Huxley above.

As various wavelengths will no longer be contracted to form our common qualia ('redness' from certain electromagnetic frequencies, etc.) because memory has lost its grip on our body (and thus environment), colour changes can occur, as well as those qualia of the other senses. Moreover, this also implies that our common sense of time may not be processed in the usual way: if, for example, we can contract fewer waves into a sensation, our present could be felt as a longer duration compared to an everyday duration of the same 'objective' time (though even that Newtonian notion is rendered obsolete by the Theory of Relativity). Bergson explicitly distinguishes *duration* from *time*, the former being the lived-in experience, the latter being the physicists' artificial time line (T_1, T_2, etc.). Ultimately, there can be no absolute standard duration as an artificial moment (T_n) cannot per se determine its experienced duration. Other creatures indubitably experience different durations to us: a fly perhaps considering our human motions preposterously slow. The psychedelic experience can allow us into these different modes of duration, providing potentially fascinating inhuman phenomenology.

Another aspect of the psychedelic experience is that of finding objects, which in sobriety are often overlooked, immensely beautiful and fascinating, especially 'organisms'. This is, in opposition to the eyes-closed entrance into the pure memory/creativity province, a contrary movement away from pure memory towards the object of which you are always a part. The nervous system is not functioning as it has evolved to function, abstracting from the environment useful extracts. Now instead, it can perceive an object with less practical bias and so really feel the fuller essence of it, also part of itself – *less extraction and more absorption*. Further still, in accord with panpsychism, as all organisms have a subjectivity (a basic consciousness) it is not impossible that one can enter into these otherwise-alien minds – diffusing into the panpsychic universe.

The divergent avenues that can be experienced under the psychedelic state are of course dependent on various factors such as substance type, the dose, the person and his/her current mindset and physiological status, the current milieu and environment, preconceptions, inculcations, and more.

In sum, the psychedelic experience is not simply abstract hallucinations caused by chemicals in the brain, but rather the diffusing of the individual consciousness into the larger reality and into alternative modes of being. *It is the ordinary everyday consciousness that is the hallucination in the sense that it is but a mere fractional-practical perspective of reality.*

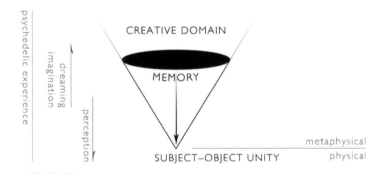

Bergson, H. (1999/1896). *Matter and Memory,* 6th edition. Zone Books, New York.

BERGSON, H. (1999/1903). *An Introduction to Metaphysics.* Hackett, Indianapolis.

Bergson, H. (1998/1907). *Creative Evolution.* Dover, New York.

Chalmers, D. J. (2010). *The Character of Consciousness.* Oxford University Press, New York.

DELEUZE, G. (1997/1966). *Bergsonism,* 4th edition. Zone Books, New York.

GAGLIANO, M. et al. (2014). Experience teaches plants to learn faster and forget slower in environments where it matters. *Oecologia,* vol. 175 (Issue 1, May). 63-72.

Hume, D. (1993/1748). *An Enquiry Concerning Human Understanding,* 2nd edition. Hackett, Indianapolis

Huxley, A. (2004/1954). *The Doors of Perception..*Vintage, Reading.

JAMES, W. (1897). *Human Immortality: Two Supposed Objections to the Doctrine*. Houghton Mifflin, Boston.

Kant, I. (1998/1781/7). *Critique of Pure Reason*. Cambridge University Press, Cambridge.

Leary, T. Metzner, R. and Alpert, R. (2008/1964). *The Psychedelic Experience*. Penguin, London.

Nagel, T. in Honderich, T. (ed.) (2010). *The Oxford Companion to Philosophy*. Oxford University Press, Oxford.

Nietzsche, F. (2008/1886). *Beyond Good and Evil*. Oxford University Press, New York

Nietzsche, F. (1998/1887). *On the Genealogy of Morality*. Hackett, Indianapolis

REID, C. R. et al. (2012). Slime mold uses an externalized spatial "memory" to navigate in complex environments. *Proceedings of the National Academy of Sciences of the United States of America*, vol. 109 (no. 43, October). 17490–17494.

Russell, B. (2007/1927). *The Analysis of Matter*. Spokesman, Nottingham.

Schopenhauer, A. (1966/1818). *The World as Will and Representation, volume 1*. Dover, New York.

Shilpp, P. A. (ed.) (1951). *Albert Einstein: Philosopher-Scientist*, 2nd edition. Tudor, New York.

Whitehead, A. N. (1997/1925). *Science and the Modern World*. Free Press, New York.

Whitehead, A. N. (1978/1929). *Process and Reality (corrected ed.)*. Free Press, New York.

EPILOGENIC CONSCIOUSNESS - A DOSE BY YET ANOTHER NAME

DAVE KING

The literature on psychedelics has left us with no shortage of names for these curious plants and compounds. These include "psychotomimetics," "hallucinogens," "entheogens," "mysticomimetics," "deleriants," and "phantasticants". This chapter follows the established tradition and offers yet another name: "epilogens". Using the model of "epilogenic consciousness", this chapter will provide an explanation for the impressive versatility of psychedelic consciousness, and draw light upon an important phenomenon: epilogenesis. The argument is summarised as follows: (1) the human system is 'aware' of every internal process, at every level, and of every interaction with external bodies from the atomic level upwards, but (2) is only 'consciously aware' of very little of that; (3) there are multiple 'domains of awareness', and (4) conscious awareness (consciousness) may expand or contract within and between domains; (5) in a state of expansion, the human may experience conscious awareness of processes that were previously autonomous and unconscious; (6) 'choice' is how we describe a decision-making process that is recognised, and may be interfered with, by conscious awareness, thus (7) a state of consciousness expansion in which ordinarily autonomic decision-making processes are consciously attended to may permit an enhancement of choice; in conclusion, (8) psychedelics sometimes act as 'choice-enhancing' catalysts of consciousness, or *epilogens*, from the Greek roots *epiloyi* (meaning choice, selection, option) and genesthai (meaning 'to bring into being').

Psychedelics have been given names that try to communicate the experience they provoke, but these experiences vary depending on who is using them, under what circumstances, with what expectations and with what degree of integration. In light of this, perhaps what is needed is a rather dull word for the plants and chemicals themselves – one that doesn't attempt to describe the states they may or may not induce – and instead build our vocabulary of the states themselves. A 'psychedelic drug' may or may not induce 'psychedelic consciousness': the user may not respond, the dose may be too low, or it may provoke a state of consciousness that fails to meet the qualifications of 'psychedelic' (for instance, a state of anxiety). Likewise, the experience may or may not be entheogenic (depending on whether or not one encounters divinity), hallucinogenic (whether there are visions), psychotomimetic (whether there is a psychotic episode), mysticomimetic (whether there is a mystical experience), or empathogenic (whether there are changes in mood and sociability).

It is perhaps not very important what the drugs are called if our interest is in the state of consciousness. If one's interest is in the drugs, rather than the consciousness (e.g. the perspectives of legislation or pharmacokinetics), it is useful to have a word that lumps disparate substances into a category. Yet, what characteristics do the drugs have in common if not the state of consciousness? We have plants, fungi, animals and synthetic chemicals that may induce psychedelic consciousness when eaten. There are some similarities pharmacologically – for instance several of the key compounds act as partial agonists at the $5-HT_{2A}$ receptor – but we also have psychedelic compounds that act selectively as NMDA antagonists, anticholinergics, kappa opioid agonists or upon the endocannabinoid system (Julian, Advokat, & Comaty, 2008). Not only is there variation between agents that may cause psychedelic consciousness, there are states which occur naturally or resultant from particular behaviours that may be considered psychedelic (e.g. fasting, sensory deprivation, hypnosis, trancework). So how can we define the substances if not by the state of consciousness, and how can we define the state of consciousness if it is subject to so much variation? No wonder we have such a difficult time reaching a consensus on what is and what is not a 'psychedelic'.

Fortunately, this chapter pays no attention to what is or what is not a 'psychedelic'. Many factors may contribute to the likelihood of a psychedelic state of consciousness, but it is the state itself that is important. Isn't it usually the state of consciousness that is important? Don't we eat to belay the uncomfortable state of consciousness that is hunger, and to bring about the pleasant states induced by the taste of the food and the satisfaction of being full? Would we as a species be so interested in sex if it didn't give such a gratifying state of consciousness? Isn't 'in love' a state of consciousness? What lengths do we not go to in our attempts to convert fear and sadness into peace and happiness?

I write the above as a preface to the following statement: there is no such thing as an *epilogen* except in reference to a particular state of consciousness. To claim that LSD is an epilogen (or an entheogen, or a hallucinogen) is misleading, but to say "the subject had an episode of epilogenic consciousness in which LSD was the epilogen", is quite acceptable.

CONSCIOUS AND NON-CONSCIOUS AWARENESS

Consciousness expands and contracts all the time – a phenomenon which Ralph Metzner has dedicated much attention to (Metzner, 2009). To understand what these terms mean, we need to consider that what we are conscious of at any one time is only a small part of what we could potentially be conscious of. Systems in the body and mind perceive and process a huge amount of information, of which, for the most part, we are not ordinarily conscious: our individual cells are involved with

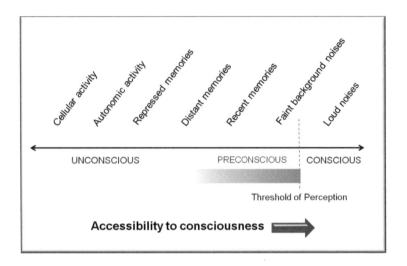

complex internal processes and chemical interactions; veins and arteries around our bodies pulse as the heart pushes blood; our subconscious mind broods on hidden conflicts and desires, affecting our moods; and so forth.

Let us imagine a spectrum of awareness within the body, as shown in FIGURE 1. At the far left of the scale is material that is highly inaccessible to consciousness; at the far right is material that is extremely accessible to consciousness. Conscious and non-conscious material is divided by the threshold of perception: when material crosses this threshold, we achieve conscious awareness of it. Material on this spectrum that approaches the threshold and which can ordinarily be brought into consciousness can be thought of as the Freudian preconscious: that which is easily

FIGURE 1: Examples of how material may fall on a spectrum of accessibility to consciousness. Material may move up and down the spectrum as it becomes more or less amenable to conscious attention. A fresh memory from a traumatic experience may move down the spectrum as it is repressed, and may move up again during a state of consciousness in which it can be re-accessed. Conscious awareness of our spiritual selves may be pulled up temporarily during a state of psychedelic consciousness, or gradually brought up the spectrum by long-term practice of meditation and related disciplines. Material may be brought into the conscious from the preconscious by bottom-up or top-down processing of attention

capable of becoming conscious (Freud, 2005). The preconscious is not a fixed domain, for material that is ordinarily not capable of becoming conscious may become so. There is continuous migration between material in the conscious, the preconscious and the unconscious.

Processes that occur in the unconscious may directly affect conscious perception. In 1910, Cheves Perky asked participants to visualise specific images (e.g. a leaf, a banana, a book) while concentrating on a fixed point on a glass screen. Images of the objects in question were projected onto the screen but were too faint to cross the threshold of perception, the result being that the images were invisible to the conscious mind. Nevertheless, the projected images influenced the ways in which the participants visualised the images (e.g. that the imagined banana was vertical rather than horizontal).

Material may be transferred into consciousness through distinct mechanisms. If a background humming slowly grew from an inaudible murmur to a deafening roar, bystanders would become consciously aware of it once the threshold of perception had been breached. This is known as bottom-up processing of attention. The threshold exists because consciousness would otherwise become overwhelmed with information unimportant to survival. If material becomes conscious because we pay attention to it, such as music in the background, conversation at the next table, our own breathing, etc., it is known as top-down processing of attention (Buschman & Miller, 2007).

In FIGURE 1 we see material approaching the threshold (i.e. that which at a certain point is called preconscious) and material that is distant from the threshold. This symbolises the ease by which that material can be transferred. A recent memory may be transferred to consciousness more easily than a distant memory; an important memory more easily than an unimportant memory; a repressed feeling less easily than one that is not. Some material may never be amenable to transfer into consciousness, other material may on rare occasions or only in particular states of mind be transferred. Most mothers have no knowledge of the sex of their child before it is born, but some have 'gut instincts' which turn out to be correct. We have no conscious awareness of tumours growing stealthily in our bodies, nor of the gradual build up of atherosclerosis, despite our immunological selves being quite aware of the problem and working tirelessly in attempt of a remedy. It is not common to retrieve perinatal memories, but some patients may do so, perhaps under hypnosis (Weiss, 2004) or with LSD-assisted psychotherapy (Grof, 2008).

Metzner (2009) identifies four domains into and from which conscious awareness may expand and contract:

- ideas, thoughts, and cognitive processes
- visual images (plus sounds, smells, tastes etc.)
- emotions, affects, feelings
- body sensations (tactile, kinaesthetic, thermal, etc.)

There may well be more domains yet to be identified. Simply put, a domain is a source of information that may be the subject of conscious attention. The spectrum, ranging from unconscious to conscious, is likely observable regardless of the domain of the material. Identifying the domain from which material is transferred is extremely important for our understanding of human experiences. Consider, for instance, near-death experiences. In these unusual states, are subjects becoming consciously aware of material from ordinarily inaccessible spiritual domains, or from ordinarily accessible sense domains? Is the light at the end of the tunnel a glimpse into the beyond, or is it the result of tunnel vision as the eyes are deprived of blood and oxygen?

THE MODEL OF CONSCIOUSNESS EXPANSION AND CONTRACTION

Expansion and contraction of consciousness is quite a banal affair. It happens all the time. Consciousness expansion occurs when the focus of conscious attention is broadened. Metzner reminds us that every morning, upon awakening, our consciousnesses are expanded: firstly, perhaps, to the cause of awakening (e.g. an alarm clock); secondly to the immediate environment: the bed, the room, our sleeping partners and so forth; and thirdly to the broader environments and social contexts in which we find ourselves. Conversely, we experience consciousness contraction every time we focus our attention on a particular event or action and in doing so lose consciousness of other things. Addictions and compulsive disorders are examples of repetitive contractions of consciousness where focus is unhealthily placed on a particular behaviour – a behaviour that is the avenue to a more pleasant (or less unpleasant) state of consciousness.

Consciousness expansion and contraction are terms that need to be placed in context. In the example above, the person waking up experiences an expansion of consciousness into the domain of sense data, but experiences a simultaneous contraction of consciousness from whatever domain we consider dreams to arise from. As we achieve acute awareness of one thing, we may lose acute awareness of another; as consciousness moves into one domain of awareness it may move away from another. The example of expansion upon awakening also allows us to describe differences in states of consciousness, which Metzner notes as being identifiable by changes in self-perception and in the experience of time and space (the Mind-Space and Time-Stream of a state) (2009). The dream-state may quite radically deviate from the confines of time, space and self-identity which are common to the waking state. Countless different states of consciousness are possible, depending on the set, setting, and triggers.

Metzner offers another analogy for contraction and expansion: imagine watching television. You may focus all your attention on the face of a particular character, which is contractive, or you may expand the focus of your attention to include the whole show, or to the room in which you're watching the television, and the knowledge that the programme is simply one of many shows and distinct from the reality of the room. You may also change the channel, which is analogous to consciousness moving to a different domain of awareness (e.g. suddenly jolting out of a very vivid daydream).

How do we quantify to what extent consciousness is contracted or expanded? Is it a matter of expanding the conscious, or the preconscious? In other words, is it expansion of what is being consciously attended to, or what can potentially be attended to in a given moment? This chapter

proposes four different types of consciousness expansion:
- acute expansion, which is the immediate state of attentive focus described in Metzner's TV analogy
- episodic expansion, characterised by a period in which the potential for acute expansion is heightened, or which significantly facilitates transference of new material into consciousness
- chronic expansion, in which there is a permanent unlocking of material which can readily be made conscious
- parallel expansion, in which domains are locked and unlocked

Chronic expansion is something achieved in all walks of life. Table 1. lists three behaviours or methods for chronically expanding consciousness in each of the four domains of awareness listed by Metzner. Each of the methods involves time, energy and practice to slowly drag material further along the spectrum shown in FIGURE 1, toward consciousness.

DOMAIN	BEHAVIOUR	MATERIAL BROUGHT INTO CONSCIOUSNESS
	Sport	Muscle control, spatial awareness
Bodily sensations: tactile, thermal, kinaesthetic, etc.	Music	Muscle control, breath control
	Yoga	Muscle control, energetic control
	Reflection	Feelings, beliefs, motivations
Emotions, affects & feelings	Therapy	Desires, inner conflicts, pain from traumas
	Loving	Love, empathy, understanding

	Optometry	Visual information
Visual images, taste, smell, noises, etc.	Wine-tasting	Subtle gustatory and olfactory information
	Perfumery	Subtle olfactory information
	Reading	Vocabulary, thoughts, ideas
Thoughts, ideas & cognitive processes	Travelling	Languages, cultural beliefs, thoughts
	Conversation	Challenging ideas and thoughts

PSYCHEDELIC CONSCIOUSNESS EXPANSION

"[Psychedelics] increase the energetic niveau in the psyche and body which leads to manifestation of otherwise latent psychological processes." (Grof, 2008)

In a state of psychedelic consciousness, there may occur heightened transference of previously non-conscious material into consciousness, which may include repressed memories, ordinarily overlooked sense data, and subtle somatic sensations. The psychedelic state may also permit phenomena including communication and interaction with a realm of spirits (Narby, 1998), vivid closed-eye visions, connection with divinity (Steindl-Rast, 2012), or out-of-body experiences (Jansen, 2004). These phenomena may turn out to be the result of a transference of material from the usually accessible domains of awareness, which would imply that these visions and out-of-body consciousnesses are happening during regular waking life but privy only to the unconscious mind. However, it is also possible that these experiences reflect the movement of consciousness into wholly different domains of awareness to those usually accessible in the waking state. As mentioned previously, this is an important distinction to make, for it will help explain the source of the phenomena.

To expand on this point further: imagine that a patient is undergoing psychedelic-assisted psychotherapy and comes across a powerful and traumatic forgotten memory. In Grof's experience, "repressed unconscious material, including early childhood memories, becomes easily available" during a session of psychedelic psychotherapy (2008). The narrative that the patient may well take from the experience is that they have re-experienced a repressed memory, and the integration of that experience may have significant therapeutic value. Recovery of genuine repressed memories is common, but there have also been cases where the memory later turns out to be false (McElroy & Keck, 1995). In the case of a genuine repressed memory, the experience derives from conscious awareness of a traumatic childhood event, but in the case of a false memory the experience may derive from the rather enigmatic domain known as 'imagination' or 'fantasy'.

Although the effect of psychedelics on unconscious material is considered non-specific, researchers are aware of factors that contribute to the content of the experience. The established notions of set, which includes the intention of the user, and setting, which pertains to the environment in which the experience occurs, describe the primary factors that affect the probability of material becoming conscious (Leary, 1992). Set describes material from the personal, internal domains, and setting describes material from the public, external domains. Imagine having a psychedelic experience in the company of people who are aggressive and untrustworthy: the setting may be violent and the set may be one of fear and discomfort. If one has a psychedelic experience in the context of focused, ongoing psychotherapy, there is a significant probability, verging on certainty, that the material made conscious will be pertinent. The less preparation and expectation there is for a psychedelic experience, the broader the scope for material able to be transferred – a potentially dangerous situation. As Grof writes, "it is necessary to structure and approach the experience in a specific way to make the emergence of unconscious material therapeutic rather than destructive" (Grof, 2008).

A psychedelic experience is thus, usually, expansive in the sense of episodic contraction/expansion. Nevertheless, during the experience the subject will undergo the usual ebbs and flows of acute contraction and expansion as his attention migrates. It may contract as he focuses on the web of tiny, interlinking grooves on his palm; it may expand as he contemplates the world-at-large. Consciousness may billow out into internal visionary domains and simultaneously withdraw from external physical domains. If he becomes briefly lost in a state of anxiety or paranoia, his consciousness will contract completely upon the discomfort and on its perceived sources.

CONSCIOUSNESS EXPANSION & CHOICE

When an ordinarily-unconscious cognitive or physical process is transferred into consciousness, it may benefit (or suffer) from being newly amenable to conscious interference. In other words, it becomes a process in which conscious choice may be exercised. Choice is a way of describing decisions that are made with conscious knowledge and attention (although I shall leave it to the neuroscientists and the philosophers to decide whether such a thing is illusory or not). I may choose to wave my hands in the air by consciously directing my body to do so, or I may unconsciously and instinctively wave them in passionate gesticulation. The less conscious the process, the less choice can be exercised in regard to it: one may easily become conscious of breathing and choose to manipulate it, but one cannot easily become conscious of digestion and choose to not digest a meal.

This enhancement of choice is parallel with the control gained from the practices listed in Table 1. A person learning a new sport may at first be lacking in skill, resultant from poor knowledge and control of the necessary muscles. With practice and attention, this knowledge is brought into consciousness on its way to procedural memory. A person who has never carefully attended to the muscles in his back may complain of general back pain, but be unable to identify precisely where. A sportsman in the same situation might be able to identify it as a cramp in the left trapezius muscle.

The various reported applications of psychedelic consciousness include the treatment of ordinarily-unconscious processes such as addictions and compulsions, allergic reactions and negative behaviour patterns. It is proposed that the mechanism for these phenomena is epilogenesis: the migration of ordinarily non-conscious material into consciousness, thereby permitting choice. Consider the following example, given by the physician Andrew Weil during his talk at the conference of the Multidisciplinary Association for Psychedelic Studies:

> "I was very allergic as a kid, in all sorts of ways. I had hay fever, I got hives in response to various things. One of the allergies I had was to cats: whenever a cat got near me my eyes would itch, my nose would run; if I touched the cat this would get much worse and if the cat licked me I would break out in hives... so I had a mindset that I was allergic to cats and didn't want them in my presence... there was a deep, ingrained defensiveness in my interactions with cats. One day, when I was twenty-eight, I took LSD with some friends. I was in a terrific space, the world was magical, everything was wonderful, and into this scene a cat bounded and jumped in my lap. I had a split second of the habitual reaction and suddenly I decided this was silly, why did I have to do this? So I started petting the cat, I started playing with the cat, I had no reaction. I haven't had a reaction to a cat since, and that was almost forty years ago."   (Weil, 2010)

Weil describes a process that is "deep" and "ingrained", from which we can surmise that it was unconscious, and how during an LSD experience he "suddenly decided" to change that process, from which we can assume that the process had been brought into consciousness. Thus, Weil was able to exercise choice where he had not previously been able. In the case of psychedelic psychotherapy being useful for the treatment of addiction (in which the addict suffers a loss of choice), one of many studies is that of the Maryland Psychiatric Research Center, published in

1970 (Pahnke, Kurland, Unger, Savage, & Grof). Hospitalised alcoholics receiving high-dose LSD treatment were significantly more likely to overcome their addiction than those receiving treatment with an active placebo. Fifty-three percent of participants in the high-dose group were considered "essentially rehabilitated" after six months, as compared to 33% percent of the low-dose group.

Curiously, under high-doses of psychedelics, it is common to feel overwhelmed and even paralysed by enhanced choice – it is not always a liberating experience. When this essay was presented at a lecture hosted by the University of Kent's Psychedelics Society, the author asked the audience to raise their hands if they had ever experienced, during a state of psychedelic consciousness, confusion as a result of forgetting how to perform a normally autonomous behaviour due to an unusual demand for conscious direction. Examples included speaking, walking, swallowing and opening a bottle of water. Around 60% of the audience confirmed that they had had such an experience.

CONCLUSION

In review of the chapter so far: we have seen how processes and information may be placed on a spectrum based on how easily they may be the subject of conscious attention, and that this material falls into one of several domains of awareness depending on its source. We also have discussed how consciousness may appear to expand or contract within and between these domains of awareness, and how material may migrate up and down the spectrum of consciousness. Finally, it has been suggested that we can consider some psychedelic experiences to permit enhanced choice by bringing ordinarily unconscious processes into consciousness. It is the author's feeling that this particular phenomenon deserves a name, for recognition of it will aid conversation and debate about the applications of consciousness expansion and may help some users of psychedelic substances to better prepare for and integrate their experiences. The term 'epilogenic consciousness' is suggested, from the Greek roots *epiloyi* (meaning choice, selection, option) and *genesthai* (meaning 'to bring into being').

Failings of this proposal include possible confusion about the circumstances under which epilogenesis occurs. The reader might feel, quite rightly, that one has very little choice over what one experiences during psychedelic consciousness. But epilogenesis doesn't explain what psychedelic consciousness is, it simply offers a description of what it can do. A significant point of critique in the proposal is that it necessarily invokes the debate of choice and free will. There are many interesting criticisms that may arise from this avenue of thought.

Buschman, T. J. & Miller, E. K. (2007). Top-Down Versus Bottom-Up Control of Attention in the Prefrontal and Posterior Parietal Cortices. *Science, 315* (5820), 1860-1862.

Freud, S. (2005). The Concept of the Unconscious. In A. Freud (Ed.), *The Essentials of Psychoanalysis* (pp. 127-190). Vintage, London.

Grof, S. (2008). *LSD Psychotherapy* (4th ed.). MAPS. Ben Lomond.

Jansen, K. (2004). *Ketamine: Dreams and Realities* (2nd ed.). MAPS, Santa Cruz.

Julian, R. M., Advokat, C. D. & Comaty, J. E. (2008). *A Primer of Drug Action.* Worth, New York.

Leary, T. M. (1992). *The Psychedelic Experience: A Manual based on the Tibetan Book of the Dead* (2nd ed.). Citadel, New York.

McElroy, S. L. & Keck, P. E. (1995). Recovered memory therapy: False memory syndrome and other complications. *Psychiatric Annals, 25* (12), 731-735.

Metzner, R. (2009). *MindSpace and TimeStream.* Regent Press, Berkeley.

Narby, J. (1998). *The Cosmic Serpent.* Orion Books, London.

Pahnke, W. N., Kurland, A. A., Unger, S., Savage, C. & Grof, S. (1970). The Experimental Use of Psychedelic (LSD) Psychotherapy. *J. Amer. Med. Assoc., 212* (1856).

Steindl-Rast, D. (2012). Psychoactive Sacraments. In Roberts, T. (Ed.), *Spiritual Growth with Entheogens: Psychoactive Sacramentals and Human Transformation* (pp. 1-5). Park Street Press, Vermont.

Weil, A. (2010, 16-April). *The Future of Psychedelic & Medical Marijuana Research.* Available at: https://vimeo.com/12057747.

Weiss, B. (2004). *Messages from the Masters* (4 ed.). Piatkus Books, London.

HARNESSING NEUROGENESIS: PSYCHEDELICS & BEYOND

SAM GANDY

The human brain is one of the most complex and amazing things in the known universe. It is made up of around 86 billion neurons, with around 86 trillion connections between these neurons. This figure is far, far vaster than the number of stars in the Milky Way galaxy or galaxies in the known universe. It is a wondrous feat of evolution, a product three and a half billion years in the making. The health of our brain is obviously very important to our happiness and well-being in life, and there are many different ways we can positively influence it. Neurogenesis and changes in neuroplasticity are receiving increasing attention for the functional role they play in the brain.

Neurogenesis is the birth of new neurons from neural stem or progenitor cells in the brain. This process is documented to take place in the brain over the adult life span of humans and other species (Spalding et al., 2013). Neurogenesis is known to decline with age, but there are a number of behavioural, environmental, pharmacological and biochemical factors that affect this process, many of which we have considerable power to influence (Nasrallah et al., 2010). Neurogenesis is also linked to changes in neuroplasticity, which is referring to changes in synapses and neural pathways in the brain, due to the effects of environment, behaviour and neural processes, although the two things can occur independently. This area is an example of an interesting and complete scientific turn around. It was long considered that the number of neurons was fixed and they did not replicate after maturity of the brain. It wasn't until the 1990s that neurogenesis was observed in the brains of humans, other primates and a number of other species that led to its widespread scientific acceptance.

Neurogenesis has been found to occur in two brain regions; the subventricular zone and the hippocampus (Nasrallah et al., 2010). The latter part of the brain plays a key role in learning and memory (Kubesova et al., 2012), and alterations in the hippocampus have been linked to a variety of cognitive pathologies such as anxiety, depression, Post Traumatic Stress Disorder (PTSD), addiction (Canales, 2013) and neurodegenerative diseases. A reduction in hippocampus volume has been observed in patients with depression and other cognitive disorders.

Increased neurogenesis can work on a number of levels. It can act via an increase in neuron proliferation, differentiation, an increase in survival rates of new neurons, or affect the maturation and integration of new neurons. Certain growth factor proteins are implicated in playing a key role in this process. Nerve growth factor (NGF) has been found to play an important role in neuroplasticity, while brain-derived neurotrophic factor (BDNF) and glial cell line-derived trophic factor (GDNF) also influence neurogenesis in the brain (Love et al., 2005).

A number of lifestyle factors have a profound influence on neurogenesis, such as exercise, diet, sleep, physiological and psychological stress, substance use, sex, environment, exposure to sunlight and meditation can all influence the process. Of these, chronic stress and excessive alcohol use are probably two of the most common and widespread factors that have a detrimental effect on this process. Cardiovascular exercise is one of the most effective ways of enhancing neurogenesis, and it boosts levels of both BDNF and GDNF.

With a few interesting exceptions, the majority of psychoactives tend to have a negative impact on neurogenesis. Of these chronic alcohol use is likely one of the most common causes of impaired hippocampal neurogenesis (Anderson et al., 2012). Alcohol intake also increases levels of the stress hormone cortisol, having a further knock-on effect (Badrick et al., 2008). Tobacco, stimulants, opiates, entactogens and some psychedelics are associated with a reduction in neurogenesis (Canales, 2013). However several agents considered drugs of abuse have been found to increase neurogenesis and produce neuro-adaptations in certain brain regions, and this may hold promise in the treatment of addiction and other conditions (Eisch & Harburg, 2006).

Psilocybin is the active compound in psychedelic mushrooms and has a long history of human use (Carhart-Harris et al., 2011). Low doses of the compound, and chronic high doses, have recently been found to increase hippocampal neurogenesis in mice, and increased the speed of their 'unlearning' of negative fear behaviour responses when compared to drug free controls, an effect that may hold promise in the treatment of Post Traumatic Stress Disorder and related pathologies (Catlow et al., 2013). Thus psilocybin may have applications for treating PTSD in humans, a syndrome characterised by highly abnormal brain function, including impaired hippocampal function (Shin et al., 2006). It may be that psilocybin, like ketamine, alters glutamatergic neurotransmission, this being linked with neurogenesis and neuroplasticity in the brain (Vollenweider & Kometer., 2010).

Research with high doses of psilocybin has found it can induce long-term, positive changes in personality and feelings of life satisfaction and well being (Griffiths et al., 2011). Openness is one of the five measures of personality, and can be significantly changed in the long-term by a single dose of psilocybin, especially if people have a mystical experience during a session (ibid.). This is of great interest, as after the age of 30 personality is thought to be generally fixed in the individual, and openness is generally thought to decline with age. This change in personality remained as strong 14 months after the session, and appears to be long-term (ibid.). Openness covers personality traits such as an appreciation for new experiences, broadness of imagination and finding value in aesthetics, emotion and curiosity, with an increased hunger for knowledge. Such traits are associated with enhanced neurogenesis and neuroplasticity. The endogenous trace amine transmitter N,N-dimethyltryptamine (DMT) is structurally very closely related to psilocybin, and its precise role in humans remains unknown; however it is hypothesised to play a role in tissue protection and regeneration (Frecska et al., 2013) and it may play a role in neurogenesis (McKenna, D., 2013, pers. comm.).

Ibogaine is a key alkaloid responsible for the psychoactivity of the root bark of *Tabernanthe iboga* and has a long history of use for spiritual and healing purposes in central Africa (Barabe, 1982). Ibogaine has been found to increase levels of GDNF in the brain (He & Ron, 2006). It is used medically for its addiction interrupting effects (Mash et al., 2000) which may be in part due to its long-term effects on GDNF expression (Messer et al., 2000) via an autoregulatory, positive feedback loop (Barak et al., 2011). The increase in GDNF expression in turn signals neurons to increase levels of mRNA, which act as biochemical blueprints in the production of further GDNF, and a single dose of ibogaine can increase GDNF expression for months. GDNF acts as a potent survival factor for a number of different dopaminergic neuronal populations in different brain regions (Boscia et al., 2009) and has been found to induce neuronal sprouting in the brain (Love et al., 2005). Ibogaine is also highly lipophilic and remains in body tissue for months, gradually being released, so further extending its influence on GDNF expression.

It appears this increase in GDNF expression is responsible for part of ibogaine's well known addiction interrupting effects (Carnicella et al., 2008). GDNF has been shown to have potent neurotrophic factor effects in both rodent and primate models of Parkinson's disease (Kirik et al., 2004) which occurs via progressive degeneration of dopaminergic neurons in the subsantia nigra on the midbrain (Lindvall & Wahlberg, 2008). GDNF does not cross the blood brain barrier (Patel & Gill, 2007), and cranial infusion has been used in the treatment of Parkinson's disease, but this is an invasive procedure and it is difficult to distribute the drug to target areas of the brain, and the side effects can be severe (Yasuhara et al., 2007).

In the future it may be possible to use compounds like ibogaine as pharmacological vectors, to increase GDNF expression in the brain in a non-invasive fashion.

Ketamine is receiving increasing attention for its antidepressant potential, particularly for the many people who are treatment resistant to other antidepressants or at a high suicide risk. Ketamine has been found to increase BDNF in rat hippocampus (Yang et al., 2013) and significantly increase plasma BDNF levels in humans (Haile et al., 2013) and it is possible that the antidepressant effects of ketamine may be related to elevated expressions of BDNF (Haile et al., 2013). It also rapidly induces growth of new synapses (Tedesco et al., 2013), and can reduce synaptic deficits associated with stress (Duman & Aghajanian 2013). Single sub anaesthetic doses are capable of inducing antidepressant effects in both human patients and in animal models within hours which may be partly due to its effects on glutamate receptors (Tedesco et al., 2013) and downstream effect on hippocampal BDNF (Yang et al., 2013). There is evidence of neurotoxicity in animal models following acute, chronic exposure to ketamine, but there are limitations to these studies and it is difficult to extrapolate these findings to humans, and further research is required (Green & Coté 2009). Chronic use of ketamine is associated with bladder damage and potential neurotoxicity, but it is possible it could form the basis of a new range of safer and more effective antidepressants.

Cannabis appears to have a positive effect on neurogenesis. Studies with a synthetic cannabinoid closely related to THC in rats found an increase in hippocampal neurogenesis, with associated anti-anxiety and antidepressant behaviour. When this process was blocked via X-irradiation, neurogenesis was blocked and the anti-anxiety behaviour was no longer observed, suggesting a link between the two (Jiang et al., 2005). Much cannabis has been bred to be high THC (responsible for the psychoactive effects) and low cannabidiol (CBD), which competes for the same receptors in the brain. Both THC and CBD have been found to act as neuroprotective antioxidants against the effects of excessive alcohol use and neuro-degeneration of the white matter of the brain (Jacobus et al., 2009). CBD is more common in indica rich strains and has been implicated with neurogenesis (Campos et al., 2013) and as a neuroprotective agent (Hampson et al., 1998). It is a molecule of increasing medical interest; as well as being a powerful antioxidant, it is an anti psychotic compound that balances out the effects of THC. CBD containing strains may be preferable with regards to brain health, and to get the most out of cannabis, vaporising and consuming it orally are the most healthy and efficient methods. The CBD levels in cannabis are down to the genetics of the plant. The cannabinoid cannabichromene (CBC) has also been demonstrated to influence neurogenesis, via enhancing the survival rates of progenitor cells as they differentiate into different types of neuron (Shinjyo & Di Marzo 2013). It is worth noting that exposure of the adolescent rat brain to THC may lead to a deficit in spatial working memory and alterations in hippocampal neuroplasticity, such as lower dendritic length and fewer synaptic connections (Rubino et al., 2009), and cannabis use by minors may well entail cognitive risk.

Plants such as the ayahuasca vine (*Banisteriopsis caapi*) and Syrian rue (*Peganum harmala*), are sources of harmine and both have a long history of human use as medicines. Harmine acts as a reversible monoamine oxidase A inhibitor and has antidepressant effects in humans. Acute administration has been found to increase BDNF levels and induce antidepressant-like effects in rat hippocampus, and may suggest a novel pharmacological target for the treatment of depression (Fortunato et al., 2009). Use of ayahuasca has been found to lead to a long-term increase in platelet 5-hydroxytryptamine (5-HT) transporters (Callaway et al., 1994) and recent neuroimaging research has shown changes in brain structure in long-term users, with associated beneficial effects, this being indicative of changes in neuroplasticity (Riba, 2013). It is these transporter sites that SSRIs act on and exert their neurogenic effects, and it may be that ayahuasca influences neurogenesis via its actions on these transporters, which may play a role in positive effects reported by users on cognitive pathologies such as anxiety, depression and addiction (Thomas et al., 2013).

Melatonin is an endogenous neurohormone produced by the pineal gland that has neuroprotective and antioxidant properties and may maintain and augment neurogenesis. It has been found to increase cell proliferation and survival in the hippocampus of aging mice (Ramírez-Rodríguez et al., 2012), and production of the hormone declines with age in both mice and humans (Ramírez-Rodríguez et al., 2012). Melatonin has also been found to enhance the survival of new neurons, and encourage growth and maturation of dendrites and lead to greater dendritic complexity and an increased volume of the granular cell layer in the hippocampus of adult mice (Dominguez-Alonso et al., 2012). The modulating effect of melatonin in neurogenesis could have important implications regarding cognitive ageing and neuropsychiatric disease (Dominguez-Alonso et al., 2012). Although melatonin levels decline with age, there are ways to increase production of the hormone, via lifestyle activities such as certain types of meditation and yoga (Tooley et al., 2000; Harinath et al., 2004) particularly when applied to late evening time, and one can supplement one's own melatonin production through diet, with certain foods such as fruits, seeds, grains and vegetables being rich sources of dietary melatonin and its precursory amino acid, 5-Hydroxytryptophan (5-HTP). Dietary melatonin appears to be far superior to taking it in the form of a synthetic supplement.

There is an ever growing body of scientific evidence to support the benefits of meditation, particularly of mindfulness meditation. It requires no belief system in order to be utilised effectively, and has been found to lead to an increase in grey matter density in a number of different brain regions, including the hippocampus, and is a powerful stress reliever (Hölzel et al., 2011). Both long-term and short-term meditation practice have been linked with changes in brain activity in regions associated with depression, anger, fear, attention and anxiety and these changes may result from physical alterations in brain structure (ibid.). Yoga may also benefit brain function, with increased hippocampal volume observed in elderly yoga practitioners (Harpiprasad et al., 2013).

The acts of learning and experiencing new things are associated with neurogenesis and enhanced neuroplasticity, so a hunger for knowledge and a life of learning and experiencing is a recipe for a healthy brain. Neurogenesis is a cutting edge frontier of research still in its infancy, and there is much that remains unknown about the implications of this process and the functional role it plays. We have the power to influence this amazing process however, and it may act to keep depression, anxiety and neurodegenerative diseases at bay, while allowing us to maintain healthy brain function into old age.

Anderson, M. L., Nokia, M. S., Govindaraju, K. P. & Shors, T. J. (2012). Moderate drinking? Alcohol consumption significantly decreases neurogenesis in the adult hippocampus. *Neuroscience, 224,* 202-209.

Badrick, E., Bobak, M., Britton, A., Kirschbaum, C., Marmot, M. & Kumari, M. (2008). The Relationship between Alcohol Consumption and Cortisol Secretion in an Aging Cohort. *The Journal of Clinical Endocrinology & Metabolism,* 93, (3), 750-757.

Barabe, P. (1982). The Religion of Iboga or the Bwiti of the Fangs. *Medicina Tropical,* 12. 251-257.

Barak, S., Ahmadiantehrani, S., Kharazia, V. & Ron, D. (2011). Positive autoregulation of GDNF levels in the ventral tegmental area mediates long-lasting inhibition of excessive alcohol consumption. *Translational Psychiatry,* 1, (12), e60.

Boscia, F., Esposito, C. L., Di Crisci, A., de Franciscis, V., Annunziato, L., et al. (2009). GDNF Selectively Induces Microglial Activation and Neuronal Survival in CA1/CA3 Hippocampal Regions Exposed to NMDA Insult through Ret/ERK Signalling. PLoS ONE, 4, (8), e6486.

Callaway, J. C., Airaksinen, M. M., McKenna, D. J., Brito, G. S. & Grob, C. S. (1994). Platelet serotonin uptake sites increased in drinkers of ayahuasca. *Psychopharmacology,* 116, (3), 385-387.

Campos, A. C., Ortega, Z., Palazuelos, J., Fogaca, M. V., Aguiar, D. C., Díaz-Alonso, Ortega-Gutiérrez, S., Vázquez-Villa, H., Moreira, F. A., Guzmán, M., Galve-Roperh, I. & Guimarães, F.S. (2013). The anxiolytic effect of cannabidiol on chronically stressed mice depends on hippocampal neurogenesis: involvement of the endocannabinoid system. *International Journal of Neuropsychopharmacology,* 16, 1407-1419.

Canales, J. J. (2013). Deficient plasticity in the hippocampus and the spiral of addiction: focus on adult neurogenesis. *Current Topics in Behavioural Neuroscience,* 15, 293-312.

Carnicella, S. & Ron, D. (2009). GDNF-a potential target to treat addiction. *Pharmacology & Therapeutics,* 122, (1), 9-18.

Cathart-Harris, R. L., Erritzoe, D., Williams, T., Stone, J. M., Reed, L. J., Colasanti, A., Tyacke, R. J., Leech, R., Malizia, A. L., Murphy, K., Hobden, P., Evans, J., Fielding, A., Wise, R.G. & Nutt, D. J. (2011). Neural correlates of the psychedelic state as determined by fMRI studies with psilocybin. *Proceedings of the National Academy of Sciences of the United States of America,* 109, 2138-2143.

Catlow, B. J., Song, S., Paredes, D. A., Kirstein, C. L. & Sanchez-Ramos, J. (2013). Effects of psilocybin on hippocampal neurogenesis and extinction of trace fear conditioning. *Experimental Brain Research,* 228, (4), 481-491.

Domínguez-Alonso, A., Ramírez-Rodríguez, G. & Benítez-King, G. (2012). Melatonin increases dendritogenesis in the hilus of hippocampal organotypic culture. *Journal of Pineal Research*, 54, (4), 427-436.

Duman, R. S. & Aghajanian, G. K. (2013). Synaptic Dysfunction in Depression: Potential Therapeutic Targets. *Upsala Journal of Medical Sciences*, 118, (1), 3-8.

Eisch, A, J. & Harburg, G.C. (2006). Opiates, psychostimulants, and adult hippocampal neurogenesis: Insights for addiction and stem cell biology. *Hippocampus*, 16, (3), 271-286.

Fortunato, J. J., Réus, G. Z., Kirsch, T. R., Stringari, R. B., Stertz, L., Kapczinski, F., Pinto J. P. Hallak, J. E., Zuardi, A. W., Crippa, J. A. & Quevedo, J. (2009). Acute harmine administration induces antidepressive-like effects and increases BDNF levels in rat hippocampus. *Progress in Neuro-Psychopharmacology & Biological Psychiatry*, 33, (8), 1425-1450.

Frecska, E., Szabo, A., Winkelman, M.J., Luna, L. E. & McKenna, D. J. (2013). A possibly sigma-1 receptor mediated role of dimethyltryptamine in tissue protection, regeneration and immunity. *Journal of Neural Transmission*, 120, (9), 1295-1303.

Green, S. M. & Coté, C. J. (2009). Ketamine and neurotoxicity: clinical perspectives and implications for emerging medicine. *Annals of Emergency Medicine*, 54, (2), 181-190.

Griffiths, R. R., Johnson, M. W., Richards, W. A., McCann, U. & Jesse, R. (2011). Psilocybin occasioned mystical-type experiences: Immediate and persisting dose-related effects. *Psychopharmacology (Berl)*, 218, 649-665.

Haile, C. N., Murrough, J. W., Iosifescu, D. V., Chang, L. C., Al Jurdi, R. K., Foulkes, A., Iqbal, S., Mahoney, J. J., De La Garza, R., Charney, D. S., Newton, T. F. & Mathew, S. J. (2013). Plasma brain derived neurotrophic factor (BDNF) and response to ketamine in treatment-resistant depression. *The International Journal of Neuropsychopharmacology*, 17, (2), 331-336.

Hampson, A. J., Grimaldi, M., Axelrod, J. & Wink, D. (1998). Cannabidiol and (-)Δ⁹ – tetrahydrocannabinol are neuroprotective antioxidants. *Proceedings of the National Academy of Sciences of the United States of America*, 95, 8268-8273.

Harinath, K., Malhotra, A. S., Pal, K., Prasad, R., Kumar, R., Kain, T. C., Rai, L. and Sawhney, R. C. (2004). Effects of Hatha Yoga and Omkar Meditation on Cardiorespiratory Performance, Psychologic Profile, and Melatonin Secretion. *The Journal of Alternative and Complementary Medicine*, 10, (2), 261-268.

Hariprassad, V. R., Varambally, S., Shivakumar, V., Kalmady, S. V., Venkatasubramanian, G. & Gangadhar, B. N. (2013). Yoga increases the volume of the hippocampus in elderly subjects. *Indian Journal of Psychiatry*, 55, (3), 394-396.

He, D. Y. & Ron, D. (2006). Autoregulation of glial cell line-derived neurotrophic factor expression: implications for the long-lasting actions of the anti-addiction drug, Ibogaine. *The FASEB Journal*, 20, 2420-2422.

Hözel, B. K., Carmody, J., Vangel, M., Congleton, C., Yerramsetti, S.M., Gard, T. & Lazar, S.M (2011). Mindfulness practice leads to increases in regional brain gray matter density. *Psychiatry Research*, 191, 36-43.

Jacobus, J., McQueeny, T., Bava, S., Schweinsburg, B. C., Frank, L. R., Yang, T. T. & Tapert, S. F. (2009). White matter integrity in adolescents with histories of marijuana use and binge drinking. *Neurotoxicology and Teratology*, 31, (6), 349-355.

Jiang, W., Zhang, Y., Xiao, L., Cleemput, J. V., Ji, S.P., Bai, G. & Zhang, X. (2005). Cannabinoids promote embryonic and adult hippocampus neurogenesis and produce anxiolytic- and antidepressant-life effects. *Journal of Clinical Investigation,* 115, 3104-3116.

Kirik, D., Georgievska, B. & Bjorklund, A (2004). Localized striatal delivery of GDNF as a treatment for Parkinson disease. *Nature Neuroscience,* 7, 105–110.

Kubesova, A., Bubenikova-Valesova, V., Mertlova, M., Palenicek, T. & Horacek, J. (2012). Impact of psychotrophic drugs on adult hippocampal neurogenesis. *Neuroscience Research,* 73, (2), 93-98.

Lindvall, O. & Wahlberg, L. U. (2008). Encapsulated cell biodelivery of GDNF: a novel clinical strategy for neuroprotection and neuroregeneration in Parkinson's disease? *Experimental Neurology,* 209, (1), 82-88.

Love, S., Plaha, P., Patel, N. K., Hotton, G. R., Brooks, D. J. & Gill, S. S. (2005). Glial cell line-derived neurotrophic factor induces neuronal sprouting in human brain. *Nature Medicine,* 11, (7), 703-704.

Mash, D. C., Kovera, C. A., Pablo, J., Tyndale, R. F., Ervin, F. D., Williams, I. C., Singleton, E. G. & Mayor, M. (2000). Ibogaine: complex pharmaco-kinetics, concerns for safety, and preliminary efficacy measures. *Annals of the New York Academy of Sciences,* 914, 394-401.

McKenna, D. (2013). Personal Communication.

Messer, C. J., Eisch, A. J., Carlezon, W. A. Jr, Whisler, K., Shen, L., Wolf, D. H., Westphal, H., Collins, F., Russell, D. S., Nestler, E. J. (2000). Role for GDNF in biochemical and behavioral adaptations to drugs of abuse. *Neuron,* 26:247-257.

Nasrallah, H. A., Hopkins, T. & Pixley, S. K. (2010). Differential effects of antipsychotic and antidepressant drugs on neurogenic regions in rats. *Brain Research,* 1354, 23-29.

Patel, N. K. & Gill, S. S. (2007). GDNF delivery for Parkinson's disease. *Acta Neurochirurgica Supplement,* 97, (Pt 2), 135-154.

Ramírez-Rodríguez, G., Vega-Rivera, N. M., Benítez-King, G., Castro-García, M. & Ortíz-López, L. (2012). Melatonin supplementation delays the decline of adult hippocampal neurogenesis during normal aging of mice. *Neuroscience Letters,* 530, (1), 53-58.

Riba, J. (2013). *Neurophysiological and neuroimaging correlates of ayahuasca effects on the human brain..* Breaking Convention, July 12-14th, London.

Rubino, T., Realini, N., Braida, D., Guidj, S., Capurro, V., Viganò, D., Guidali, C., Pinter, M. Sala, M., Bartesaghi, R. & Parolaro, D. (2009). Changes in hippocampal morphology and neuroplasticity induced by adolescent THC treatment are associated with cognitive impairment in adulthood. *Hippocampus,* 19, (8), 763-772.

Shin, L. M., Rauch, S. L. & Pitman, R. K. (2006). Amygdala, Medial Prefrontal Cortex, and Hippocampal Function in PTSD. *Annals of the New York Academy of Sciences;* 1071, 67-79.

Shinjyo, N. & Di Marzo, V. (2013). The effect of cannabichromene on adult neural stem/ progenitor cells. *Neurochemistry International,* 63, (5), 432-437.

Spalding, K. L., Bergmann, O., Alkass, K., Bernard, S., Salehpour, M., Huttner, H. B., Boström, E., Westerlund, I., Vial, C., Buchholz, B. A., Possnert, G., Mash, D. C., Druid, H., Vollenweider, F. X. & Kometer, M. (2010). Opinion: The neurobiology of psychedelic drugs: implications for the treatment of mood disorders. *Nature Reviews Neuroscience,* 11, 642-651.

Spalding, K. L., Bergmann, O., Alkass, K., Bernard, S., Salehpour, M., Huttner, H. B., Boström, E., Westerlund, I., Vial, C., Buchholz, B. A., Possnert, G., Mash, D. C., Druid, H., Frisén, J.

(2013). Dynamics of hippocampal neurogenesis in adult humans. *Cell*, 153, (6), 1219-1227.

Tedesco, V., Ravagnani, C., Bertoglio, D. & Chiamulera, C. (2013). Acute ketamine-induced neuroplasticity: ribosomal protein S6 phosphorylation expression in drug addiction-related brain areas. *Neuroreport*, 24, (7), 388-393.

Thomas, G., Lucas, P., Capler, N.R., Tupper, K.W. & Martin, G. (2013). Ayahuasca-assisted therapy for addiction: results from a preliminary observational study in Canada. *Current Drug Abuse Reviews*, 6, (1), 30-42.

Tooley, G.A., Armstrong, S.M., Norman, T.R. & Sali, A. (2000). Acute increases in night-time plasma melatonin levels following a period of meditation. *Biological Psychology*, 53, (1), 69-78.

Yang, C., Hu, Y.-M., Zhou, Z.-Q., Zhang, G.-F. & Yang, J.-J. (2013). Acute administration of ketamine in rats increases hippocampal BDNF and mTOR levels during forced swimming test. *Upsala Journal of Medical Sciences*, 118, 3-8.

Yasuhara, T., Shingo, T. & Date, I. (2007). Glial cell line-derived neurotrophic factor (GDNF) therapy for Parkinson's disease. *Acta Medica Okayama*, 61, (2), 51-56.

IN SEARCH OF THE PHILOSOPHER'S STONE

MARIA PAPASPYROU

> *"The unconscious is the mother of consciousness."*
>
> C. G. Jung (1968)

In the following pages unfolds the narrative of Maya's healing journey on an intrapsychic and interrelational level. Maya's process of transformation unfolded in the context of her psychotherapy, and was further facilitated by her own personal resolve to use entheogens for therapeutic purposes. In our therapeutic work we firmly established the importance of integrating her experiences in the realms of altered consciousness.

As a practitioner involved with the 'technologies' of the psyche, I cannot ignore the enormous therapeutic potential of psychedelic agents, and hold the view that they can be of immense contribution to psychotherapy. I have encountered their creative and powerful therapeutic potential by working as a sitter in festival settings with projects like Kosmicare, as well as in my professional practice where a handful of people over the years have brought forward the medicinal value of such experiences. In working with the numinous character of these experiences I have found the Jungian framework of analytic psychology to be an incredibly powerful aiding lens, through which to examine and integrate the psychedelic experience.

Maya was a young woman in her mid-thirties that came to see me at a point of intense familial conflict, and in particular sought to address her difficulties in her relationship with her mother, which for 20 years was caught in alienation and disconnection. For two decades they had remained stuck in a cycle of anger, resentment, and guilt, threatened at the thought of real intimacy. These patterns culminated to a point of intense and disturbing relational discordia, that led to a complete breakdown of their communications. They ceased all contact, and denied each other's existence, along with their need for each other. During this relational breakdown of a year and a half, Maya's therapeutic engagement and explorations, along with three separate experiences with three separate entheogens, gradually provided the catalysts that unlocked their relationship from its fixed conflictual narratives.

Entheogens have the capacity to very definitively facilitate a process of deintegration and reintegration of the self. The process of death and rebirth, one of the fundamental laws of nature, is an inescapable cycle we all experience throughout our lifetime in various forms and intensities. Transformation during an entheogenic journey takes place in a rhythmic process of taking the self apart and putting it back together, in evolutionary service. Chaos gives birth to new order of greater complexity than before (Papaspyrou, 2014). This process is in line with the alchemical maxim of *solve et coagula,* which means 'dissolve and coagulate', i.e. something must be broken down before it can be built back up. The early alchemists saw chaos as the pool of infinite possibility, the primordial creative space. That chaos is our unconscious and the new order is its assimilation into our consciousness. According to Jung, "the descent into the depths always seems to precede the ascent" (Jung, 1968b).

Jung distinguished between the personal and the collective unconscious (Jung, 1968c) He described the personal unconscious as a superficial layer made up of contents, once conscious, that have retreated into the unconscious by being forgotten or repressed. He saw the collective unconscious as a deeper layer, with contents that have never been in consciousness or individually acquired. For Jung this was a universal and impersonal layer, filled with timeless primordial images, the archetypes (ibid).

Archetypes are core elements of Jungian psychology and enormously relevant when studying the psychedelic experience. These archetypal messengers express themselves primarily in metaphors that mediate between unconscious depths and consciousness, with the aim of restoring wholeness of the Self (in this article, self with a lower 's' is used to denote the 'ego', the centre of consciousness, and Self with a capital 'S' refers to the psychic totality of conscious and unconscious). Conscious will by itself can not always unite a personality (Jung, 1968d), for the formation of the Self, primordial powers of great depth need to awaken and intervene.

The archetypal psyche is a powerful space in the collective unconscious, and reveals itself in spontaneous symbolic forms. The symbols that emerge from the collective layers bring into consciousness thoughts, intuitions, and affects, deeply buried within oneself. Their meaning can only be inferred with an 'as if' quality. These symbols, through reaching and being integrated into consciousness, allow us to overcome an initial situation on a higher level (Jung, 1968e) leading to the ultimate human goal of individuation, which is wholeness and unity of the Self. "The underlying, primary psychic reality is so inconceivably complex that it can be grasped only at the farthest reach of intuition, and then but very dimly. That is why it needs symbols" (Jung, 1968f).

Jung was the first to extensively psychologise alchemy. He saw the alchemical opus as a reflection of the process of individuation, and the search for the philosopher's stone as a process of seeking inner transformation (Robertson, 2009). For Jung the true purpose of the alchemical opus was the development of the alchemist's soul.

Alchemy was based on a series of cyclic operations where every stage would be taken apart and then put back together, until eventually lead would turn into gold, and the object would become the philosopher's stone. The alchemists believed that they were simply speeding up the natural processes that took place within the bowels of the earth (ibid.). This echoes the potential of entheogens to act as catalysts, speeding up the evolution of our individual and collective consciousness, through thinning the veil that separates conscious and unconscious mind, allowing these to unite, and facilitating conditions for a greater degree of integration.

The alchemical process of transformation was a three-stage process from chaos to full life (ibid.). It started from the chaotic dark space of the *nigredo*, associated with the black colour. In psychological terms this is the stage of suffering and darkness (ibid.). If the alchemist could successfully navigate the darkness of nigredo, this would give way to the *albedo*, associated with the white colour. In psychological terms this is the stage where one emerges from the dark night of the soul with a new understanding (ibid.). If the work continued successfully on a material and a spiritual level, the *rubedo* would emerge, associated with the red colour. In psychological terms, in this stage the new understandings are slowly integrated. According to Jung, a mental union marked the completion of the nigredo, a union of mind and body marked the completion of the albedo, and a union of mind, body and ultimate reality marked the completion of the rubedo (ibid.).

These alchemical stages of transformation have provided an in-depth framework for understanding the transformative processes that were initiated through the therapeutic process and the integration of Maya's entheogenic experiences. In the journeys that follow one can hear the unconscious speaking the language of the psyche.

THE NIGREDO PHASE

The nigredo phase reached its crescendo when all communications between Maya and her mother had ceased. Four months into her therapeutic journey Maya had an entheogenic experience that captured in a symbolic form the very essence of our work up until that point. She accessed the concentrated insights of this healing phase by the use of changa, and then integrated these through our work together.

Changa is a herbal mixture of *Banisteriopsis caapi* leaves and a natural extract of DMT, that create a powerful and short acting psychoactive blend. Due to its short acting nature, changa has been often used as a quick 'psychedelic hit', neglecting its immense insight potential as an entheogen, when used with attention to the set and setting, and with the focusing aid of intention.

Two visions from this journey elucidated the psychological tension that was the basis of Maya's nigredo phase.

Vision one brought to life a wooden sculpture of two fused entities. Their fusion was the source of painful friction. Maya witnessed what looked like their growth spurs resulting in awkward movements, and they were stuck in a push-pull dynamic that seemed physically intolerable.

Vision two took her to a room with a cot and a baby in it. The baby was fast growing until the room could no longer fit it, as it started growing through doors, windows, and walls.

In combination these two visions, through symbolism, pointed at the necessity of Maya detaching herself from her 'origins'. The first vision, in clear and simple symbolic representation, depicted in an embodied form the need for separateness, in order to allow space for Maya's growth and development. The child motif appearing in the second vision relates to the individuation process. The child as an archetype holds a futurity within it (Jung, 1968g). Its symbolism speaks of the anticipation of a future development, and signals a future change of personality. It is a mediator between the present and the inherent future, signifying how fragile the psychic possibility of wholeness can be. Unconscious communications can transcend linear time as "the unconscious psyche is not only immensely old; it is also capable of growing into an equally remote future" (Jung, 1968a).

The nigredo phase is characterised by chaotic unconsciousness. It is a time of inner turmoil, confusion, and uncertainty, when our vital energies retreat into the unconscious. In this space we come to face our 'shadow' through a psychic descent into our own underworld. In Jungian analytic psychology the 'shadow' refers to the unconscious, unknown or denied, aspects of oneself. This is also a time of immense value, as it is through this chaos that the philosopher's stone can ever truly emerge.

THE ALBEDO PHASE

The catalyst that facilitated Maya's transition to the albedo phase was the integration of a second entheogenic journey, about ten months into her therapeutic process, with lysergic acid diethylamide, more commonly known as LSD. We have known of LSD's psychedelic properties since Albert Hofmann's 'bicycle trip' in 1943. LSD's Akashic Field (László, 2004) is loaded by its history in the 1960s and 1970s and that legacy is carried forward by both mainstream and counter cultural misconceptions and misrepresentations. However a closer look at early psychedelic research reveals LSD to be a substance of immense therapeutic and psycho-spiritual value, if used with care and attention to the conditions of set and setting (Grof, 1975).

> Maya experienced a discomfort in her body and found a space where she could delve in and address it. An image emerged, the stone sculpture of a broken goddess. The broken goddess, the wounded mother, the aged woman all flashed before her soul in this image. She felt an immense emotional opening and her heart was flooded with love. She described what felt like her ego-structures melting away.

> Vision two followed up and it was a visual journey that narrated the story of her mother's soul. It told Maya through images that it is a very old soul that has over many lifetimes been trapped in a series of terrible deaths. It showed her an early time where the first death was in a dessert and birds were eating away the defeated, decaying flesh, a later death in the dark Middle Ages, under an emblem of authority, being condemned to extinction, a whole series of unspeakable endings.

Maya was also told that this soul, neither male nor female, is in her mother's body today, and that Maya has some responsibility for helping it transcend this Karmic obstacle. Maya entered her heart and in its core she found only love; all other layers she could now recognise as her protection. Maya, for the first time in a very long time, felt peaceful in her love for her mother as a fellow soul, free from the restraining dynamics of their material reality, and the inflexible strings of their mother-daughter relationship.

Theistic ideas are strongly associated with parental images. "The concept of the Great Mother belongs to the fields of comparative religion [and mythology] and it embraces widely varying types of mother-goddess" (Jung, 1968h), derivatives of the mother archetype. In search of meaning for Maya's vision, we entered the realm of the vulnerable goddesses (Bolen, 1984), and there we unravelled and explored the myth of Demeter and Persephone.

Demeter, a maternal archetype, goddess of harvest, grain, and earth's fertility, sank into depression and grief when Hades, lord of the underworld, abducted her daughter Persephone. In response to Demeter's grief the seasons halted, earth became infertile, and famine threatened to destroy humankind. In the face of such destruction, the gods intervened and Demeter and Persephone were reunited, with the compromise of Persephone spending two thirds of each year in the upper world, and a third in the Underworld, and fertility and growth on earth was restored. The myth became the basis for the Eleusinian mysteries, the most important rituals of ancient Greece, where initiates through undisclosed sacred rites, that some scholars believed to include the use of entheogens (Hofmann et al, 2008), experienced the renewal of life after death, mirroring Persephone's annual return from the underworld.

The myth echoed Maya's vision image of the broken goddess in the form of Demeter, as well as the second vision of her mother's metempsychosis, the transmigration of her soul through different body existences and reincarnations with the continuity of Karma, reflected in the death-rebirth ritual that was enacted in the Eleusinian mysteries.

The notion of metempsychosis might be a dubious one in our Western culture, but time is a different construct when it comes to the language of the soul, "whereas we think in periods of years, the unconscious thinks and lives in terms of millennia" (Jung, 1968a).

The Demeter archetype embodies a dual nature (Bolen, 1984), Demeter before the abduction, symbol of a nurturing, giving, and fertile mother-goddess, and Demeter after the abduction, a depressed, destructive, and withholding maternal archetype. Persephone, through leaving her origins, is transformed from Kore (maiden) to queen of the underworld. As queen of the underworld, reigning over the kingdom of the dead and guiding the living through it, Persephone has come to her own Self. Immersing into the underworld denotes the necessity for Persephone to enter the deeper layers of her self, where the personal and collective unconscious reside, in order to transform. Persephone, in her Kore and Queen duality, is an active archetype for entheogenic journeys, representing our ability to move back and forth, mediating between the ego-based reality of the real world, and the unconscious or archetypal reality of the psyche. The Persephone archetype is also of a daughter that is too close to her mother to develop an independent sense of self (ibid.). Separation is important, as Maya's earlier changa vision had revealed, and reconciliation can only be realised after she has grown into her individuated Self.

The myth provided a metaphor that allowed Maya to recognise and consciously relate to the active archetype of Demeter within her mother, and recognise and work through the active archetype of Persephone within herself. Through her work Maya was able to witness and acknowledge the possibility for herself and her mother reflected in Demeter and Persephone to grow through their suffering, and recognise that their familial narrative had become stuck at the 'abduction level'.

The hermaphroditic element of Maya's mother's soul in her second vision, neither male nor female, was another meaningful symbol at this stage of her inner alchemical transformation. The join of the opposites is the perfect symbol for what has been accomplished during the albedo phase, which is the union of the strongest and most striking opposites

(Jung, 1967g). The hermaphrodite is a "subduer of conflicts and a bringer of healing" (ibid.). The bisexual primordial being symbolises the unity of the personality, the coming to one's whole Self, where conscious and unconscious integrate. In this journey Maya's anger, rage, and distrust, were making contact with the hidden subterranean layers of her love, need, and hope. This journey provided a deep emotional and visceral breakthrough that was a powerful facilitator in Maya's sessions.

In the albedo phase, sparks of new consciousness emerge from the unconscious that transcend ego boundaries. Something tries to emerge from the collective unconscious layers, which gives it a numinous quality. Its emergence releases one from black and white positions, and more gradients are starting to emerge. The ability to hold the opposites long enough, will give rise to new truths (Robertson, 2009). "The alchemist has reached the final part of the albedo stage, having passed through the darkness of the nigredo, and emerged with an early glimpse of the numinous light that will eventually form the philosopher's stone..." (ibid.). The final stage of the work depends on loosening identification with the archetypes. Identification with these powerful forces can be psychologically destructive (Jung, 1968i; Stein, 2006; Robertson, 2009). It was now time for Maya to find her way home to the human world, bearing the gifts of the vital energies of the imaginal world.

THE RUBEDO PHASE

Maya's transition to the final alchemical phase was well under way and further facilitated by an ayahuasca journey a year and two months into our work. Ayahuasca is a powerful psychoactive brew of mixed *Banisteriosis caapi* vine and DMT-containing shrubs. Ayahuasca has had a very long history of shamanic use for medicinal, spiritual, and healing purposes.

Maya was on the first cup and thoughts floated through that put her childhood in order. Maya felt that her personal unconscious had thoroughly unlocked and granted her pristine clarity. Her mother entered the scene, and the spirit of the vine took Maya in her heart and showed her, her mother's love for her. She witnessed a vibrant field full of potential. The vine guided her through their relationship. It showed her on one side and her mother on the other and in between a black maze that appeared as an obstacle. Maya asked the vine about this obstacle, and it informed her it was her father's death. It showed her how his death had been an active wound in her relationship with her mother, and how grief had drifted them apart. Maya was advised to visit her mother on her father's imminent death anniversary and share with her what she had learned.

For Jung, in psychological terms, the final stage of rubedo marks the union of mind, body, and ultimate reality. Unable to travel on her father's 21st death anniversary, Maya made contact with her mother on the day, a year and a half after their last shared and destructive contact, and exchanged memories over her father's life and death. Maya was astonished to discover from that contact how little she knew about his final days, and how much was left unsaid between her and her mother since his departure from their physical reality. A while later Maya was able to travel and share with her mother her inner and outer journeys of transformation through her entheogenic and psychotherapeutic explorations and integrations. Her mother confirmed Maya's encounters in the realms of altered consciousness, and the detailed insights from her childhood the vine had granted her, were now grounded between them.

The work that entheogenic journeys facilitate is not miraculous, although its does not fall far short. But it takes work, effort, and a commitment to work with the openings that these intelligent agents and our incredibly wise unconscious facilitate. These are openings that reach completion when supported by integration; otherwise they easily sink back into the unconscious. Maya continued her therapeutic work for another year, and while the real world has its own rhythms and operates on many levels, she had by that point developed sufficient ego strength to maintain her sense of self while being in relationship with her mother. For Maya, reading this account of her experiences and therapeutic journey, and considering whether to grant permission for these to be published, is yet another layer of bringing her rubedo phase to completion, allowing her integrated insights to reach a union of mind, body, and ultimate reality.

In the nigredo stage we must lose our essential nature, the philosopher's stone, in the albedo stage we find it again, and during the rubedo stage we are tasked with bringing it back into the 'real' world (Robertson, 2009).

In entheogenic journeys the underlying psychic reality unravels. Entheogenic use brings the psyche upon numinous and mythic dimensions. The mythopoetic function of the unconscious is based on archaic mythological thought forms. The myths and symbols we bring back from these deep soul journeys are part of our psychic life, imbued with vital meaning. Our task is in translating the archaic speech of vision into a meaningful narrative of our earthly reality.

Such experiences are not without their dangers for they are also the very matrix of psychoses. Numinous encounters have a dark side, as powerful as their light, and "archetypes can profoundly disturb [and possess] consciousness" (Stein, 2006). To remain stuck in numinous lands equals becoming assimilated to the unconscious and to be possessed or rely defensively on archetypes can be very destructive on a personal and a collective level (ibid.). In these journeys we are treading in liminal spaces. The liminal borderlands between conscious and unconscious, are as full of creative potential as they are of disastrous ruin. Here is the stuff of madness, spiritual realisation and artistic inspiration. Set and setting, with particular emphasis on our ego strength, is what determines our ability to withstand and navigate the up-rise of unconscious contents. To integrate our journeys is to carry forward the responsibilities that such endeavours come with.

Experiences of this kind, when adequately integrated, have the potential to provide us with developmental turning points, and can be great aids to individuation, life's ultimate developmental goal. They offer the potential to widen our consciousness through the union of conscious and unconscious, what Jung called the "transcendent function" (Jung,

1969). Unconscious materials are abundantly available, but valueless unless one can creatively extract from them something meaningful to support their integration to our conscious self. The unconscious is reality in potential, and entheogenic agents are real catalyst for the emergence of this potential.

The war on drugs has jeopardised the rightful place of entheogens as agents of transformation within the healing spheres. Research has been halted for 40 years and only recently a professional movement seems to have been coming together. Promisingly, the first training schemes in psychedelic psychotherapy and psychedelic studies are starting to appear in the Unites States, aimed at equipping researchers and therapists for working with these substances within legal research settings. One can hope for the future that we can create societies that embrace the transformational potential of the various consciousness states, and support ways to include these for the unfolding of the psycho-spiritual potential of our individual and collective selves.

This is a version based on a previously published article in *JTR*, Vol. 7 (1).

Bolen, J. S. (1984). *Goddesses in Everywoman: Powerful Archetypes in Women's Lives*. Harper Paperbacks.

Grof, S. (1975). *Realms of the Human Unconscious: Observations from LSD Research*. Viking Press, New York.

Hofmann, A., Wasson, G. R., Ruck, C. A. P. (2008). *The Road to Eleusis: Unveiling the Secret of the Mysteries*. 30th Ann. Ed. North Atlantic Books.

Jung, C. G. (1968a). Conscious, Unconscious, and Individuation. *The Collected Works of C. G. Jung*: Vol. 9 Part 1. Second Edition. Routledge, London .

Jung, C. G. (1968b). Archetypes of the Collective Unconscious. *The Collected Works of C. G. Jung*: Vol. 9 Part 1. Second Edition. Routledge, London.

Jung, C. G. (1968c). The Concept of the Collective Unconscious. *The Collected Works of C. G. Jung*: Vol. 9 Part 1. 2nd Ed. Routledge, London.

Jung, C. G. (1968d). The Phenomenology of Spirit in Fairytales. *The Collected Works of C. G. Jung*: Vol. 9 Part 1. 2nd Ed. Routledge, London.

Jung, C. G.. (1968e). Archetypes of the Collective Unconscious. *The Collected Works of C. G. Jung*: Vol. 9 Part 1. 2nd Ed. Routledge, London.

Jung, C. G. (1968f). The Practical Use of Dream-Analysis. *The Collected Works of C. G. Jung*: Vol. 16 Part 1. 2nd Ed. Routledge, London.

Jung, C. G. (1968g). The Psychology of the Child Archetype. *The Collected Works of C. G. Jung*: Vol. 9 Part 1. 2nd Ed. Routledge, London.

Jung, C. G. (1968h). On the Concept of the Archetype. *The Collected Works of C. G. Jung*: Vol. 9 Part 1. 2nd Ed. Routledge, London.

Jung, C. G. (1968i). A study in the process of individuation. *The Collected Works of C. G. Jung*: Vol. 9 Part 1. Second Edition. Routledge, London.

Jung, C. G.(1969). The Transcendent Function. *The Collected Works of C. G. Jung*: Vol. 8 Second Edition. Princeton University Press.

László, E. (2004). *Science and the Akashic Field: An Integral Theory of Everything.* Inner Traditions.

Papaspyrou, M. (2014). Femtheogens: The Synergy of Scared Spheres. *Psychedelic Press UK.* 2014 (5), 29-46.

Robertson, R. (2009). *Indra's Net: Alchemy and Chaos Theory as Models for Transformation.* Quest Books.

Stein, M. (2006). Importance of Numinous Experience in Alchemy of Individuation. *The Idea of the Numinous: Contemporary Jungian and Psychoanalytic Perspectives.* Casement, A., & Tracey, D., (eds.) Routledge, London.

THE ECSTATIC HISTORY OF MDMA: FROM RAVING HIGHS TO SAVING LIVES

BEN SESSA

The drug MDMA has a controversial history; from its roots as an agent to assist psychotherapy in the 1970s and 1980s, to its wide scale popularity as the recreational drug Ecstasy in the rave scene of the 1990s, to its contemporary re-emergence as a potential tool in psychiatry for treating Post Traumatic Stress Disorder. The debate about the relative risks, safety and potential medical uses of MDMA is steeped in political issues that threaten to undermine the basic principles of evidence-based medicine. MDMA is frequently labelled with an exaggerated harm profile by the popular media, which mirrors an established political agenda, rather than following an evidence-based realistic appreciation of its relative benefits as well as its risks. If we are to truly examine whether MDMA has a place in modern medicine we must reflect on the political interferences and apply the same rules of rigorous methodology that we would with any other proposed novel pharmacological agent. Failure to do this threatens withholding a potentially relatively safe and efficacious medicine from those patients for whom MDMA could be of benefit.

INTRODUCTION: WHAT IS MDMA AND ECSTASY?

Is Ecstasy the saviour of vinyl records and the raison d'être of dance floor culture? Or is it a gateway drug to harder substances and the cause of individuals' if not the whole of society's problems? Could MDMA be an essential tool for personal spiritual development and improved community cohesion? Or even an important clinical medicine to improve the outcomes of seriously ill psychiatric patients?

MDMA might be all or none of these things. It is certainly a misunderstood compound, with erroneous assumptions and preconceptions from both the popular press and from within medicine. In order to understand its future place in medicine and culture we must explore the history and development of MDMA; from its early uses as a medicine to assist psychotherapy, to its popular use as a recreational drug in the rave culture, right through to the current renaissance of interest in clinical MDMA. Many consider MDMA "the perfect drug for Post Traumatic Stress Disorder (PTSD) psychotherapy", with an increasing number of contemporary studies around the world employing MDMA-assisted psychotherapy as a tool to treat this insidiously increasing psychiatric disorder (Sessa, 2006). We must examine the benefits and risks of the drug, take a realistic analytical approach and question the reported negative media bias.

MDMA has taken an interesting journey from clinics to raves and back. If we are to imagine its viable future we need proper evidence-based facts. After a 40-year hiatus there is now a renaissance underway regarding the medical profession's approach to psychedelic drug therapy. MDMA and psilocybin are emerging as drugs more likely than their old cousin LSD, to lead the way in heralding this new form of medical treatment; psychedelic drug-assisted psychotherapy for the treatment of anxiety-based disorders.

Firstly we must put MDMA in context against other similar drugs, those we classify as psychedelic.

WHAT ARE PSYCHEDELIC DRUGS?

The 'classical' psychedelic drugs include LSD, psilocybin, DMT and mescaline; all of which are 5-HT_{2A} receptor partial agonists. Another group of psychedelics, the NMDA antagonist dissociatives, include ketamine, PCP and DXM. Other dissociative anaesthetics are Kappa-Opioid agonists, which include ibogaine and *Salvia divinorum*. MDMA comes from a group commonly known as the entactogens; all potent serotonin, dopamine and noradrenaline receptor agonists. MDMA has chemically-related cousins, MDA, 2CB, 2CI and 2CT7, which share a similar structure.

In terms of chemical structure most psychedelic drugs fall into one of two groups, either tryptamines or phenethylamines. Tryptamines include dimethyltryptamine, psilocybin, LSD and ibogaine. And the phenethylamines include mescaline and MDMA. The root molecules tryptamine and phenethylamine occur naturally in the brain, leading some researchers to postulate there are endogenous psychedelic chemicals released into the brain during non-drug mystical and spiritual experiences (Strassman, 2001).

THE CROSS-CULTURAL AND ARCHAIC USES OF HALLUCINOGENS

Psychedelic drugs have been used by non-Western cultures for millennia. The use of peyote, ibogaine, ayahuasca, psilocybin, *Amanita muscaria* mushrooms and other naturally occurring plants and fungi, including cannabis, can be traced back to ancient cultural history and has been at the birth of most religions (Hofmann, Schultes & Rätsch, 1996). But MDMA is a relatively new drug.

MDMA

MDMA (3,4-Methylenedioxymethamphetamine) is also known as the recreational drug Ecstasy. It is relatively short acting, lasting two to five hours, making it more clinically manageable than LSD. It is also less perceptually distorting than LSD, with a predictable subjective effect that is almost always pleasurable. It has particular empathogenic and entactogenic qualities and it is physiologically safe in proposed therapeutic applications.

MDMA'S HISTORICAL TIMELINE

MDMA was first synthesised and patented by the German chemical company Merck in 1912 as a precursor chemical for an unconnected chemical synthesis. It was never marketed or distributed by Merck, but surfaced again in the 1950s as part of the US government's MK Ultra project. In the 1960s drug culture MDMA played a surprisingly small role. 1967, the first Summer of Love, embraced the widespread use of LSD, with only a mild interest in MDA, a close cousin of MDMA. In 1967 a graduate student first introduced the chemist Alexander 'Sasha' Shulgin to MDMA. He thereafter began using the drug with his close friends calling it "my low-cal martini". Shulgin developed a new synthesis technique and introduced MDMA to the psychotherapist Leo Zeff, who had been a psychedelic therapist with LSD until it was banned in 1966. Zeff was pleased to find MDMA, then still legal, could be used as a substitute for LSD in therapy. He spread the word and hundreds of West Coast psychotherapists began experimenting with MDMA (then called 'Adam') (Zeff, 2004). Therapists found the drug particularly useful for couples therapy and in the early 1980s a few anecdotal clinical case series were subsequently published (Greer & Tolbert, 1986).

But by then 'Adam', which was also then being called 'Empathy', was increasingly common in nightclubs in Dallas and being used by followers of Bhagwan 'Osho' Rajneesh, who moved with his followers from Oregon, USA to the Island of Ibiza. In 1985 the DEA, concerned about the growing recreational use, put MDMA into Schedule 1. Some MDMA psychotherapists in the USA challenged the DEA and a judge ruled the drug should not be Schedule 1 but Schedule 3, as there were recognised clinical uses. However the DEA overruled the judgment and MDMA was placed permanently in Schedule 1, disallowing any research. In the wake of this legal issue, Rick Doblin founded the Multidisciplinary Association for Psychedelic Studies (MAPS) to campaign for a resumption of MDMA research (MAPS n.d).

Growing in popularity, dealers relabelled MDMA as Ecstasy and in 1987 three young teenagers, Nicky Holloway, Danny Rampling and Paul Oakenfold, holidaying in Ibiza, discovered Ecstasy and brought it back to the UK where they started the London nightclub *Shoom*. Rave was born. The second Summer of Love, 1988, saw Ecstasy use spread throughout the UK and in 1992, at the famous Castle Morton Rave, travellers of the 1970s free festival scene met the emerging young ravers head on. By 1993 the rave scene was mainstream and the UK band The Shamen took *Ebeneezer Goode* to number one. With a number of high profile deaths of young people and the subsequent Criminal Justice Bill banning raves, the illegal drug Ecstasy became public enemy number one in the eyes of the general public and the media.

MDMA RESEARCH DEVELOPMENTS

Between 1988 and 1993 there was a relaxation of laws in Switzerland allowing a small group of psychotherapists to deliver psycholytic therapy with MDMA and LSD – some of whom continued in underground practice, whilst elsewhere political restrictions slowed research progress (Sessa & Meckel Fischer, 2011). It was not until 2010 that Michael Mithoefer, working with MAPS, would finally publish the world's first randomised controlled study testing MDMA-assisted psychotherapy against placebo. Since then there has been a massive renaissance of psychedelic research (Sessa, 2012).

IS MDMA DANGEROUS? IS MDMA SAFE?

The answer to both these questions is: it depends. Many modern activities can be either dangerous or safe, including bungee-jumping, playing squash, horse-riding, taking prescribed drugs, drinking beer, having sex and crossing the road. They can all be dangerous but can also all be relatively safe if done moderately and with care and attention. We must look at MDMA in the same way, with a realistic risk-versus-benefit analysis.

THE RISKS OF MDMA

Figures suggest around 40 people a year die in the UK in part from MDMA (Office for National Statistics, 2010). Deaths are primarily from hyperthermia or the idiosyncratic reaction hyponatraemia. More common though less severe side effects of taking MDMA are that of bruxism and oro-bucco-facial acute dystonia (Sessa, 2007). But this figure of 40 deaths per year includes deaths involving other drugs mixed with MDMA, frequently opiates and alcohol. Only 7% of deaths involve MDMA alone (Schifano et al., 2006). This equates to just three deaths annually – from an estimated 150 million doses of ecstasy consumed by over a million people each year. Furthermore, this relates to the recreational use of Ecstasy, not clinical MDMA. A recreational Ecstasy pill may contain anything, including ketamine, caffeine, BZP, heroin, brick dust, dog deworming tablets, fish tank oxygenation tablets, methamphetamine, mephedrone or opiates. Ten years ago the quantity of MDMA in Ecstasy tablets was around 60mg per pill, but more recent data from the Netherlands suggests it could now be much higher (Cole et al., 2002; Koning, 2013).

THE BENEFITS OF MDMA: PTSD

MDMA has been researched for its potential role in treating Post Traumatic Stress Disorder (PTSD), a severe anxiety-based psychiatric disorder, with a high level of treatment-resistance and high rates of mortality from suicide. PTSD usually follows a catastrophic life event such as a car accident, natural disaster, episode of combat in war or a single discrete violent assault. But it may also be the result of repeated, less life-threatening but equally frightening traumatic events, such as chronic child sexual abuse. A lack of integration of the sensory experience of the episode, incorrect storage of memories and a disruption to REM sleep results in re-experiencing phenomena of the traumatic event; presented as flashbacks and nightmares. Hyper-arousal and hyper-vigilance of cues that remind the sufferer of the index trauma make PTSD a difficult and complex disorder to treat. PTSD is associated with high levels of drug misuse, alcohol dependency and high rates of self-harm and suicide. The lifetime prevalence is between 6% and 10% and half of sufferers develop the chronic disorder (Kessler et al., 2005). PTSD is increasingly associated with post-combat trauma; previously known as 'shell shock' and 'battle fatigue'. Given the recent wars in Iraq and Afghanistan there are high levels of sufferers of post-combat PTSD in the United States and the UK.

Traditional treatment involves multiple pharmacotherapies, with drugs used symptomatically rather than to affect an endurable remission. Treatment resistance is common because the severity of the traumatic memories frequently overwhelms the patient and makes traditional trauma-focused therapy difficult. When asked to recount the trauma as part of their therapy, patients 'dissociate', re-experiencing the trauma acutely; often withdrawing from treatment and fleeing towards drug or alcohol misuse and self-harm.

WHY MDMA IS SO RIGHT FOR PSYCHOTHERAPY FOR PTSD

Because of its unique receptor profile MDMA has an effect on positive mood and creative thinking, gentle stimulation to motivate for treatment, gentle relaxation to treat hyper-vigilance and hyper-arousal and crucially, it vastly increases a sense of empathy and bonding both between the therapist and the patient and allows the patient to experience feelings of empathy towards their abuser. In combination these effects have great value at keeping a patient engaged in psychotherapy for PTSD (Sessa, 2012).

RECENT TREATMENTS WITH MDMA IN PSYCHOTHERAPY FOR PTSD

The results of the first pilot study for PTSD using MDMA-assisted psychotherapy have been positive. Compared to 15% of placebo subjects, 85% of patients with treatment-resistant PTSD no longer met the criteria for the PTSD diagnosis after a single course of MDMA-assisted psychotherapy, which involved taking MDMA three times during a 16-week course of weekly psychotherapy sessions. These results were sustained four years later (Mithoefer, 2012). The results are significantly better than the current PTSD treatments with Cognitive Behavioural Therapy (CBT) and antidepressants, so clearly this study needs to be replicated. Further studies are now underway or being planned to look at the role for MDMA-assisted psychotherapy for treating PTSD in the USA, Australia, Canada, Israel and the UK.

NO INTERVENTION IN MEDICINE IS 100% RISK FREE

Every doctor is aware on a daily basis of the risk-versus-benefit argument. All interventions in medicine carry some degree of risk – whether this be a sticking-plaster, cancer chemotherapy, complex surgery or, indeed, administering MDMA to patients with PTSD. Before a doctor decides on which treatment to use and whether it is safe for his or her patient, they always consider this basic analysis; if the benefit can be shown to outweigh the risk then the treatment can be justified. When one looks at MDMA with the risk-versus-benefit argument, it is clear that giving a moderate dose of MDMA on a limited number of occasions, in a controlled clinical setting, and under medical supervision, with careful follow-up, is a risk worth taking, given the potential benefit; that of potentially relieving otherwise unremitting PTSD.

POLITICAL ISSUES

But when it comes to MDMA there is not a level playing field. In 2009 the UK's Professor David Nutt was asked to provide a report about Ecstasy/MDMA on behalf of the British government. He undertook an unbiased exploration of the relative risks of the drug, consulting with experts in the fields of medicine, neuroscience and the criminal justice service. Nutt concluded that Ecstasy/MDMA was less risky than these professions had initially thought and that it ought to be moved from Class A into Class B to better reflect the evidence-based relative harms and safety. But this suggestion did not fit with the then Labour government's political agenda, which did not want to appear 'soft on drugs'; so the Home Secretary disregarded Nutt's advice, saying to change the drug's class would be to "send the wrong message". Professor Nutt, whose allegiance lies with that of objective evidence-based scientific proof, felt quite rightly that, surely, the only message ought to be the truth!

He spoke out against the Home Secretary's decision and was subsequently sacked from his position as drugs advisor to the government. Nutt went on to form an independent scientific body (the ISCD), which continues to this day to publish and disseminate the results of studies into drugs without any biased influence from governments.

HOW POLITICS INFLUENCES MEDICAL RESEARCH

There has always been a strong anti-Ecstasy lobby in the UK and United States, with powerful political allegiances, which at times directly impacts on medical research. This has held back research into MDMA as a treatment for PTSD (Sessa & Nutt, 2007). Political agendas influenced the vocal neurotoxicity arguments against MDMA that were common in the late 1990s. Politically motivated government-backed groups of researchers publicised exaggerated neuro-cognitive deficits based on animal models, that did not relate at all well to the pattern and dosage of MDMA as proposed for clinical therapy to treat PTSD. For example, studies used to suggest the risk of neurotoxicity in humans included those by Hatzidimitriou and Ricaurte (1999), who gave primates the equivalent of a human taking 40 Ecstasy tablets in four days and Sabol (1996), who gave rats the human equivalent of 20 tablets of Ecstasy every 12 hours until the animal had consumed 160 tablets (Hatzidimitriou & Ricaurte, 1999). But the biggest, and sadly one of the most erroneously influential, flawed studies of MDMA research is that of George Ricaurte at Johns Hopkins University, USA. Commissioned by the US government, Ricaurte's study allegedly demonstrated severe neurotoxicity in primates given MDMA (Ricaurte et al., 2002). But it was later discovered that the bottles in the testing lab were incorrectly labelled and the animals were actually given the highly toxic drug methamphetamine instead of MDMA (Ricaurte et al., 2003). Although Ricaurte subsequently retracted his study from *Science* and offered an apology, the damage was done and the subsequent government and media campaigns of "holes in the brain" still remain as many governments' political message; their smoking gun against MDMA.

Interestingly the neurotoxicity argument has become less prominent in recent years, countered by the mounting epidemiological evidence against it. The UK general public has been taking MDMA in high quantities for 25 years and we have simply not seen these patients in our psychiatric clinics or hospitals at anything like the levels that were predicted back in 1988. MDMA is no longer a public health issue. There do remain, however, some largely discredited researchers who continue to suggest that even a single dose of MDMA can cause permanent brain damage (Parrott et al, 2011). But in my opinion such approaches merely betray such researchers' backgrounds as being non-clinical. As described earlier, a clinician is not looking for a fictional 100% safe drug. Rather they appreciate the risk-versus-benefit argument and accept that, even if it does carry some small risk, it could still be beneficial if used cautiously with the patient.

A MORE JUDICIOUS APPROACH TO DRUG CLASSIFICATION; FOR THE SAKE OF MDMA RESEARCH

Repeated commentators have been encouraging governments to look at a wide range of illegal drugs with a realistic risk-versus-benefit analysis. Nutt's study in *The Lancet* ranked MDMA as number 18 out of 20 in terms of its relative harm (Nutt et al., 2010). However these evidence-based arguments have often not been appreciated by either the government of the day nor the general public. All medical interventions carry some degree of risk and MDMA is no exception. But in the absence of evidence-based politics, doctors and research scientists in this field continue to struggle against a negatively biased, media-driven political agenda in which the MDMA baby is thrown out with the bathwater (Nutt, King & Nichols, 2013). Discounting any possible therapeutic role for MDMA in Post Traumatic Stress Disorder is denying patients the opportunity for a potentially life-saving treatment. There are 4,400 deaths from suicide each year in the UK and tens of thousands of sufferers of this chronic condition. If MDMA could be seen to reduce the burden of morbidity and mortality in PTSD then it is more accurate to think of this drug as a saver of lives rather than a substance to be feared.

SUMMARY

Research with MDMA and other psychedelic substances remains controversial. This is an area of medicine fraught with strong feelings and passionate arguments on both sides of the debate. The association between MDMA and the historically abused recreational drug Ecstasy is unhelpful. If we are to strive to have MDMA recognised as a viable tool for the treatment of PTSD we must carefully steer a course away from ingrained stereotypes about recreational psychedelic drug-use and concentrate on clear evidence-based data emerging from robust methodologically-sound scientific studies. If, under these conditions, MDMA is shown to be effective and safe then we owe it to our patients with treatment-resistant Post Traumatic Stress Disorder to research this area with great haste.

Declarations of Interest: None

Cole, J. C. et al. (2002). The content of ecstasy tablets: implications for the study of their long-term effects. *Addiction*, 97, 1531-1536.

Greer, G.R. & Tolbert, R. (1986). Subjective Reports of the Effects of MDMA in a Clinical Setting. *Journal of Psychoactive Drugs*, 1986, Vol. 18(4):319-327.

Hatzidimitriou, G. et al. (1999). Altered Serotonin Innervation Patterns in the Forebrain of Monkeys Treated with (6)3,4-Methylenedioxymethamphetamine Seven Years Previously: Factors Influencing Abnormal Recovery. *Journal of Neuroscience*, June 15, 1999, 19(12):5096–5107.

Hofmann, A., Schultes, R. E., Rätsch, C. (1996). *Plants of the Gods: Their Sacred Healing and Hallucinogenic Powers*. Healing Arts Press, Rochester, Vermont.

Kessler, R. C., Chiu, W. T., Demler, O., et al. (2005). Prevalence, severity, and comorbidity of 12- month DSMIV disorders in the National Comorbidity Survey Replication. *Arch Gen Psychiatry* 2005;62:617–27.

Zeff, L. (2004). *The Secret Chief Revealed*. MAPS.

Mithoefer, M. C., et al. (2012). Durability of improvement in post-traumatic stress disorder symptoms and absence of harmful effects or drug dependency after 3,4-methylenedioxymethamphetamine-assisted psychotherapy: a prospective long-term follow-up study. *J Psychopharmacology*. 2013 Jan; 27(1):28-39.

Nutt, D. J., King, L. A., Nichols, D.E. (2013). Effects of Schedule I drug laws on neuroscience research and treatment innovation. *Nat Rev Neuroscience*. 2013 Jun 12.

Nutt, D. J. et al. (2010) Drug harms in the UK: a multicriteria decision analysis. *The Lancet*. 6 November 2010. Vol. 376, Issue 9752, Pages 1558-1565.

Office for National Statistics (2010). *Deaths related to drug poisoning in England and Wales, 2010*

Parrott A. C., et al. (2011). MDMA and methamphetamine: some paradoxical negative and positive mood changes in an acute dose laboratory study. *Psychopharmacology* 215: 527-536.

Koning, R., Amsterdam – Personal communication. July 2013.

Ricaurte, G. A., et al. (2002). Severe Dopaminergic Neurotoxicity in Primates After a Common Recreational Dose Regimen of MDMA ("Ecstasy"). *Science* 297: 2260-3.

Ricaurte George A, et al (2003) Retraction. *Science* 301: 1429.

Sabol, K. E., et al. (1996) Methylenedioxymethamphetamine-Induced Serotonin Deficits Are Followed by Partial Recovery Over a 52-Week Period. Part I: Synaptosomal Uptake and Tissue Concentrations. *Journal of Pharmacology and Experimental Therapeutics*. 276:846-854, 1996

Schifano, F., et al. (2006) Ecstasy (MDMA, MDA, MDEA, MBDB) consumption, seizures, related offences, prices, dosage levels and deaths in the UK (1994 – 2003). *Journal of Psychopharmacology* 20 (3) 456 – 463.

Sessa, B., Nutt, D. J. (2007). MDMA, politics and medical research: have we thrown the baby out with the bathwater? *Journal of Psychopharmacology* 21: 787-791

Sessa, B. & Meckel Fischer, F. (2011). Underground LSD, MDMA and 2-CB-assisted Individual and Group Psychotherapy in Zurich: Outcomes, Implications and Commentary. *Journal of Psychopharmacology / Journal of Independent Scientific Committee on Drugs*.

Sessa, B. (2006) From Sacred Plants to Psychotherapy: The History and Re-Emergence of Psychedelics in Medicine. *Quarterly Journal of Mental Health*, Vol. 1, # 2, 2006

Sessa, B. (2007). Is there a role for MDMA Psychotherapy in the UK? *Journal of Psychopharmacology.* Vol 21; 220-221

Sessa, B. (2012). Could MDMA be useful in the treatment of PTSD? *Progress in Neurology and Psychiatry.* Vol. 15:6. 4–7

Sessa, B. (2012). Shaping the renaissance of psychedelic research. *The Lancet* - 21 July 2012. Vol. 380: 9838, 200-201.

Strassman, R. (2001). *DMT: The spirit molecule: A doctor's revolutionary research into the biology of near-death and mystical experiences.* Park Street Press, Rochester, Vermont.

MDMA–ASSISTED PSYCHOTHERAPY: AN UPDATE OF COMPLETED AND ONGOING RESEARCH

MICHAEL MITHOEFER

WHY STUDY MDMA FOR PTSD?

Most existing treatments for PTSD are based on revisiting and reprocessing traumatic experiences in a therapeutic setting. The most widely recognised treatment methods, Prolonged Exposure (PE) (Foa et al., 2009; Ursano et al., 2004; Benedek et al., 2009), Cognitive Processing Therapy (CPT) (Resick and Schnicke, 1992) and Eye Movement Desensitization and Reprocessing (EMDR) (Hogberg et al., 2008) are effective for approximately half the patients who undertake them, leaving a large percentage who cannot tolerate or do not respond adequately to these methods (Bradley et al., 2005). Response may be blocked by either overwhelming anxiety or emotional numbing; opposite extremes on the continuum of emotional arousal that are both common features of PTSD (Foa et al., 2009). Ogden and others have observed that, on a continuum from hyperarousal to hypoarousal, there is an "optimal arousal zone" or "window of tolerance" in which therapeutic processing can occur (Ogden et al., 2006).

MDMA was employed as a therapeutic adjunct soon after its rediscovery and prior to its being placed in Schedule 1 (Shulgin and Shulgin, 1991, Stolaroff, 2004). Early investigations suggested that MDMA was a useful therapeutic adjunct (Greer and Tolbert, 1986). MDMA, administered in a therapeutic setting, appears to have the potential to modulate arousal and thereby allow access to this optimal arousal zone for a period of several hours.

This postulated effect is consistent with neuroimaging results by a number of investigators. Gamma et al. found decreased activity in the left amygdala and increased activity in the ventromedial prefrontal cortex and the ventral anterior cingulate following MDMA administration (Gamma et al., 2000). Bedi et al. reported decreased activity in the amygdala in response to angry faces (Bedi et al., 2009), and Carhart-Harris et al. found that MDMA decreased negative affect in response to negative autobiographical memories without significantly decreasing the vividness of memory recall. This was accompanied by decreased activity in the left anterior temporal lobe, an area with dense connections to the amygdala. (Carhart-Harris et al., 2014). Separate fMRI studies of subjects with PTSD have demonstrated increased activity in the amygdala and decreased activity in the mPFC (Nutt and Malizia, 2004; Lanius et al., 2011). A pilot study of fMRI in response to neutral and trauma scripts is currently underway as part of our ongoing study of MDMA-assisted psychotherapy for PTSD in South Carolina and other such studies are under development elsewhere.

An additional obstacle for people with severe PTSD is that they may have difficulty trusting and forming an effective working alliance with a therapist. The increase in oxytocin levels following MDMA administration likely plays a role in enhancing therapeutic alliance. (Thompson et al., 2007; Dumont et al., 2009; Hysek et al., 2012; Bedi et al., 2014).

MDMA TREATMENT STUDIES

Several studies have been designed to test the hypothesis that MDMA, in conjunction with psychotherapy, can be safely administered to carefully screened individuals with PTSD, and will lead to improvement in PTSD symptoms as measured by well validated scales administered by blinded independent raters. Data from completed and ongoing Phase II clinical trials provides preliminary support for this hypothesis. All these studies to date have been sponsored by the private, nonprofit Multidisciplinary Association for Psychedelic Studies (MAPS). Because the Merck Pharmaceutical patent on MDMA expired many years ago and MAPS has intentionally avoided creating grounds for a use patent, there is no proprietary competition connected with this research (Freudenmann et al., 2006).

Bouso et al. in Spain started the first Phase II clinical trial of MDMA-assisted psychotherapy for PTSD, however this study was terminated prematurely in 2002 for political reasons (Bouso et al., 2008). In Charleston, South Carolina we completed the first Phase II MDMA trial in September 2008. Subsequently, Oehen and colleagues completed a similar study in Switzerland (Oehen et al., 2013), and there are now additional studies being conducted in South Carolina, Colorado, Israel and Vancouver, with other studies in development in England and Australia.

The safety and efficacy of ±3,4-methylenedioxy-methamphetamine-assisted psychotherapy in subjects with chronic, treatment-resistant posttraumatic stress disorder: the first randomised controlled pilot study (Mithoefer et al., 2011; Mithoefer et al., 2013). **Charleston, South Carolina 2004 -2008:**

This is a double-blind placebo controlled trial of 20 people with PTSD, mostly resulting from childhood sexual abuse or rape, who had not responded adequately to treatment with both medication and psychotherapy. The average duration of PTSD before study enrolment was more than 19 years. Sixty percent of the participants received MDMA on two or three occasions in conjunction with psychotherapy;

40% received inactive placebo with the same psychotherapy. Later, after the blind was broken, those who received therapy with placebo could enrol in a crossover trial of open label MDMA-assisted therapy.

After careful medical and psychiatric screening and two preparatory sessions with the male/female psychotherapy team, MDMA or placebo was administered during eight-hour psychotherapy sessions a month apart. Following each psychotherapy session participants spent the night in the clinic with an attendant on duty, and met with the therapy team the following morning before leaving. They then had daily phone contact with the therapists for a week and weekly follow up sessions for the next three weeks.

THE THERAPEUTIC METHOD

The method used in the study sessions is our adaptation of psychedelic and MDMA-assisted psychotherapy developed by Stan and Christina Grof, Ralph Metzner, Leo Zeff, George Greer and Requa Tolbert and others (Greer and Tolbert, 1998; Grof, 2001; Metzner and Adamson, 2001; Stolaroff, 2004). The approach is largely nondirective, aimed at supporting the individual's own emerging experience. Preparatory sessions are aimed at developing a therapeutic mindset and setting. Participants are encouraged to trust that their innate wisdom and impulse to heal will determine the direction of their therapeutic process more effectively than any predetermined agenda. While interaction with the therapists is an important element of the treatment, their primary role is to support the unfolding of this process rather than to direct it. Although this approach is by nature less structured and more flexible than most manualised treatments, we have detailed its essential elements in a manual available at: http://www.maps.org/research/mdma/MDMA-Assisted_Psychotherapy_Treatment_Manual_Version_6_FINAL.pdf (Mithoefer et al., 2013).

MDMA, administered in this setting, appears to deepen access both to affirming, joyful experiences and to difficult, painful ones. For people with PTSD, MDMA sessions usually include periods of painful

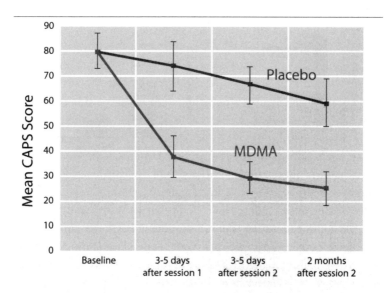

emotional processing during which a safe, supportive setting and an intention to explore rather than avoid the pain can be important in assuring a therapeutic outcome. Several participants said, "I don't know why they call this Ecstasy!", nevertheless their experience was beneficial because they were able to face the pain without being overwhelmed. MDMA did not make trauma processing easy; it made meaningful processing possible in a way that it had not been without MDMA.

Another observation from our research is that the therapeutic process catalysed by MDMA is not limited to the 4 or 5 hours of maximum MDMA effect. The process continues to unfold in the hours, days and weeks following the session. This often includes pleasurable feelings and useful insights. At other times it may bring waves of painful emotion as well as challenges to integrating new perspectives into daily life and relationships. For this reason follow-up sessions with the therapists play an important role in maximising therapeutic benefit and minimising the possibility of adverse effects.

FIGURE 1: The results of the Clinician Administered PTSD Scale (CAPS)

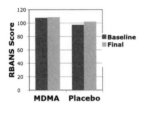
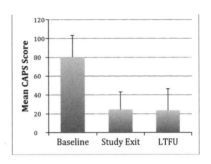

RESULTS

The results of this first study showed significant improvement in PTSD symptoms in both groups, those who were randomised to psychotherapy with inactive placebo as well as those randomised to MDMA-assisted psychotherapy, however the improvement was much larger in the MDMA group. FIGURE 1 shows the results of the Clinician Administered PTSD Scale (CAPS), our primary outcome measure and the gold standard for measuring PTSD symptom severity in PTSD research.

Clinical response, defined as > 30% reduction in CAPS scores, occurred in 25% of the placebo (psychotherapy only) group and in 83% of the MDMA-assisted psychotherapy group.

STAGE 2, OPEN LABEL CROSSOVER

After completing the double blind stage of the protocol, seven of the eight people who had been randomised to psychotherapy with placebo elected to participate in the open label crossover treatment in which they received two or three sessions of MDMA in conjunction with the same psychotherapy, followed by essentially the same schedule of integration sessions. In this group, the mean CAPS score dropped from 65.6 after therapy with placebo to 33.9 after MDMA-assisted psychotherapy ($p < 0.05$). In this group the clinical response after MDMA was 100%.

FIGURE 2: Mean RBANS Score (left) **FIGURE 3:** LTFU Mean CAPS Score (right)

NEUROCOGNITIVE MEASURES

Because research in recreational Ecstasy users has raised questions about possible effects on memory, the independent raters did comprehensive prospective neuropsychological testing before and after MDMA and placebo using The Repeatable Battery for the Assessment of Neuropsychological Status (RBANS), The Paced Auditory Serial Addition Task (PASAT) and The Rey-Osterrieth Complex Figure Test (RCFT). There were no significant group differences on any of the cognitive measures at baseline, nor any significant differences in any of the major index scores on between-group analysis two months after the second MDMA or placebo session. This is best illustrated by the RBANS total score illustrated in FIGURE 2.

LONG-TERM FOLLOW-UP (LTFU)

In our initial study the final outcome measures were administered two months after the last treatment. Subsequently, we conducted a long-term follow-up of these study participants in which we asked them to complete a LTFU questionnaire and to repeat the CAPS and other measures with the same independent rater (Mithoefer et al., 2013). Because this follow-up occurred more than a year after the last subject completed the study, the average duration since MDMA treatment was 45 months (range 17–74 months). Of the 19 people who had received MDMA in the original study, 16 agreed to do the repeat CAPS. Two of them proved to have relapsed to CAPS scores of > 50, the original cutoff for enrolment, indicating that improvement was not sustained for 12.5% of this sample. However, in the group as a whole, the mean CAPS score was 23, unchanged from the two-month score and 56 points lower than their mean baseline score (FIGURE 3). On the questionnaire all 19 subjects reported benefit and none reported harm. The three who did not complete the CAPS for various reasons did report sustained benefit according to the questionnaire, which suggests that they had not relapsed. Nevertheless, since the questionnaire is not a validated measure of PTSD symptoms, one could assume that these three individuals would have relapsed to high CAPS scores had they taken the CAPS. If this were the case, it would still mean that the majority (74%) of these previously treatment-resistant patients maintained sustained recovery from PTSD three and a half years after treatment.

This result is noteworthy not only because it is much longer than the usual follow-up period for psychiatric treatment studies, but also because it argues against the possibility that improvement in the original study was due to placebo effect from the difficulty maintaining an effective double blind with an obviously psychoactive compound. Additional evidence for a true treatment effect comes from the participants' comments on the LTFU questionnaire describing a coherent therapeutic process that, by its nature, would be expected to lead to psychological healing and growth:

"The MDMA provided a dialogue with myself I am not often able to have, and there is the long-term effect of an increased sense of well being."

ONGOING STUDIES: THERAPIST TRAINING

In order for all MAPS-sponsored studies to adhere to the same treatment method, we have developed a therapist training programme for research therapist teams that uses the manual and discussion of video recordings from study sessions. We also believe it is desirable for therapists doing this research to have had their own experience as a client in MDMA-assisted psychotherapy. For this reason we sought and obtained FDA, DEA and IRB approval for a Phase I clinical trial studying psychological measures in normal volunteers receiving MDMA in the same therapeutic setting used in PTSD studies, with enrolment limited to therapists planning to work on MDMA clinical trials who have already completed a MAPS-sponsored training in the therapeutic method. Thus far, seven therapists have been enrolled in this study and report that the experience has been an important addition to their training as MDMA research therapists (Halberstadt, 2014).

A RANDOMISED, TRIPLE-BLIND, PHASE II PILOT STUDY COMPARING THREE DIFFERENT DOSES OF MDMA IN CONJUNCTION WITH MANUALISED PSYCHOTHERAPY IN 24 VETERANS, FIREFIGHTERS, AND POLICE OFFICERS WITH CHRONIC PTSD

This study is similar in design to our first study, except that we are testing three different doses of MDMA to see if a low dose will serve as an active placebo to make it more difficult for participants and therapists to guess to which condition they were randomised (Mithoefer et al., 2012). The CAPS is still the primary outcome measure, but we have added additional measures in an attempt to capture some of the benefits study participants have described that go beyond improvement in PTSD symptoms as measured by the CAPS. The additional measures are: The Beck Depression Inventory (BDI-II), Global Assessment of Functioning (GAF), Posttraumatic Growth Inventory (PTGI), Pittsburgh Sleep Quality Index (PSQI), States of Consciousness Questionnaire (SOCQ) and the NEO Personality Inventory (NEO). Preliminary results are showing clinically significant improvements in the full and medium dose groups but not in the low dose group.

STUDIES AT OTHER SITES

Studies of MDMA-assisted psychotherapy for PTSD are under way in Israel, Colorado and Vancouver, and MDMA/PTSD studies are in development in Australia and England (Doblin, 2011). With this continued progress in Phase II trials we expect to proceed to Phase III trials that could lead to approval of MDMA as a clinical treatment delivered in specialised treatment centres designed to provide a proper set and setting for clinical use.

QUOTES FROM STUDY PARTICIPANTS

Many of our study participants have spoken eloquently about the nature of the therapeutic process, so I will close with some of their descriptions:

"The first times were like opening a door, making you aware that there even were doors to open. The higher dose was about knowing you could make the choice to go through the door and experience the feelings. It helped me remember stuff that I don't know if I ever would have remembered."

"It's helped me in so many ways... It feels like it's gradually rewiring my brain... It feels like the MDMA sessions were the crack in the ice, because the trauma was so solid before that."

"It feels almost like the inner healer or the MDMA is like a maid doing spring cleaning. It's as if you thought you were cleaning before but when you got to things you didn't really want to deal with you'd just stick them in the attic. If you're going to clean the house you can't skip the stuff in the attic."

"I keep getting the message from the medicine, 'trust me'. When I try to think it doesn't work out, but when I just let the waves of fear and anxiety come up it feels like the medicine is going in and getting them, bringing them up, and then they dissipate."

"The medicine just brought me a folder. I'm sitting at this big desk in a comfortable chair and the medicine goes and then rematerialises in physical form bringing me the next thing – this is a folder with my service record. It says I need to review it and talk to you about it from the beginning so it can be properly filed."

"I don't think I would have survived another year. It's like night and day for me compared to other methods of therapy. Without MDMA I didn't even know where I needed to go. Maybe one of the things the drug does is let your mind relax and get out of the way because the mind is so protective about the injury."

"I see huge white doors with beautiful white glass, so huge and heavy, but a master has engineered them so you can open them with one hand. It's only without the fear that the doors are so light. How interesting! If I go up to them with all the fears it makes me weak. I'm taking those fears out of different parts of my body, looking at them and saying 'it's ok but I'm leaving you here.' The fear served me well at one time, but not now for going through these doors."

"I'm a huge pile of fertiliser composting and turning into beautiful rich soil. It's a perfect time to have rain. I'm a converter, I'm the earth, I am. Leaves, rain, even acid rain hit me, and I have a powerful ecosystem, all can be absorbed. What we're doing here is turning compost."

"I realise I'm not trying to break through anything. It has to be softly opening. With the medicine nothing felt forced. I know I'm going to have to feel the feelings and there's still fear that the grief will be overwhelming, and I know feelings are unpredictable and the currents can be swirly, but yesterday when I put my toe in it felt so wonderful to feel. I remember every detail, it's a pristine, pristine image."

"I have respect for my emotions now (rather than fear of them). What's most comforting is knowing now I can handle difficult feelings without being overwhelmed. I realise feeling the fear and anger is not nearly as big a deal as I thought it would be."

ACKNOWLEDGMENTS

I wish to thank Lisa Jerome, Ph.D. for her assistance with references and her helpful suggestions about the manuscript.

BEDI, G., CECCHI, G. A., FERNANDEZ SLEZAK, D., CARRILLO, F., SIGMAN, M. & DE WIT, H. (2014). A Window into the Intoxicated Mind? Speech as an Index of Psychoactive Drug Effects. *Neuropsychopharmacology.*

BEDI, G., PHAN, K. L., ANGSTADT, M. & DE WIT, H. (2009). Effects of MDMA on sociability and neural response to social threat and social reward. *Psychopharmacology (Berl)*, 207, 73-83.

BOUSO, J. C., DOBLIN, R., FARRE, M., ALCAZAR, M. A. & GOMEZ-JARABO, G. (2008). MDMA-assisted psychotherapy using low doses in a small sample of women with chronic posttraumatic stress disorder. *J Psychoactive Drugs*, 40, 225-36.

BRADLEY, R., GREENE, J., RUSS, E., DUTRA, L. & WESTEN, D. (2005). A multidimensional meta-analysis of psychotherapy for PTSD. *Am J Psychiatry*, 162, 214-27.

CARHART-HARRIS, R. L., WALL, M. B., ERRITZOE, D., KAELEN, M., FERGUSON, B., DE MEER, I., TANNER, M., BLOOMFIELD, M., WILLIAMS, T. M., BOLSTRIDGE, M., STEWART, L., MORGAN, C. J., NEWBOULD, R. D., FEILDING, A., CURRAN, H. V. & NUTT, D. J. (2014). The effect of acutely administered MDMA on subjective and BOLD-fMRI responses to favourite and worst autobiographical memories. *Int J Neuropsychopharmacol*, 17, 527-40.

DOBLIN, R. (2011). MDMA Research update. *MAPS Bulletin*, 21, 2-4.

DUMONT, G. J., SWEEP, F. C., VAN DER STEEN, R., HERMSEN, R., DONDERS, A. R., TOUW, D. J., VAN GERVEN, J. M., BUITELAAR, J. K. & VERKES, R. J. (2009). Increased oxytocin concentrations and prosocial feelings in humans after ecstasy (3,4-methylenedioxymethamphetamine) administration. *Soc Neurosci*, 4, 359-66.

FOA, E. B., KEANE, T. M., FRIEDMAN, M. J. & COHEN, J. A. (2009). *Effective Treatments for PTSD, Practice Guidelines from the International Society for Traumatic Stress Studies*. Guilford Press, New York.

FREUDENMANN, R. W., OXLER, F. & BERNSCHNEIDER-REIF, S. (2006). The origin of MDMA (ecstasy) revisited: the true story reconstructed from the original documents. *Addiction*, 101, 1241-5.

FRYE, C. G., WARDLE, M. C., NORMAN, G. J. & DE WIT, H. (2014). MDMA decreases the effects of simulated social rejection. *Pharmacol Biochem Behav*, 117, 1-6.

GAMMA, A., BUCK, A., BERTHOLD, T., LIECHTI, M. E. & VOLLENWEIDER, F. X. (2000). 3,4-Methylenedioxymethamphetamine (MDMA) modulates cortical and limbic brain activity as measured by [H(2)(15)O]-PET in healthy humans. *Neuropsychopharmacology*, 23, 388-95.

GREER, G. & TOLBERT, R. (1986). Subjective reports of the effects of MDMA in a clinical setting. *J Psychoactive Drugs*, 18, 319-27.

GREER, G. R. & TOLBERT, R. (1998). A method of conducting therapeutic sessions with MDMA. *J Psychoactive Drugs*, 30, 371-379.

Grof, S. (2001). *LSD Psychotherapy: 4th Edition*, Ben Lomond, CA, Multidisciplinary Association for Psychedelic Studies.

HALBERSTADT, N. (2014). MDMA-Assisted Psychotherapy, A View from Both Sides of the Couch. *MAPS Bulletin*, XXIV, 4-6.

HARRIS, D. S., BAGGOTT, M., MENDELSON, J. H., MENDELSON, J. E. & JONES, R. T. (2002). Subjective and hormonal effects of 3,4-methylenedioxymethamphetamine (MDMA) in humans. *Psychopharmacology (Berl)*, 162, 396-405.

HOGBERG, G., PAGANI, M., SUNDIN, O., SOARES, J., ABERG-WISTEDT, A., TARNELL, B. & HALLSTROM, T. (2008). Treatment of post-traumatic stress disorder with eye movement desensitization and reprocessing: outcome is stable in 35-month follow-up. *Psychiatry Res,* 159, 101-8.

LANIUS, R. A., BLUHM, R. L. & FREWEN, P. A. (2011). How understanding the neurobiology of complex post-traumatic stress disorder can inform clinical practice: a social cognitive and affective neuroscience approach. *Acta Psychiatr Scand,* 124, 331-48.

METZNER, R. & ADAMSON, S. (2001). Using MDMA in healing, psychotherapy and spiritual practice. Holland, J. (ed.) *Ecstasy, A Complete Guide: A Comprehensive Look at the Risks and Benefits of MDMA.* Rochester VT: Inner Traditions.

MITHOEFER, M. C., WAGNER, M. T., MITHOEFER, A. T., JEROME, L. & DOBLIN, R. (2011). The safety and efficacy of {+/-}3,4-methylenedioxymethamphetamine-assisted psychotherapy in subjects with chronic, treatment-resistant posttraumatic stress disorder: the first randomized controlled pilot study. *J Psychopharmacol,* 25, 439-52.

MITHOEFER, M. C., WAGNER, M. T., MITHOEFER, A. T., JEROME, L., MARTIN, S. F., YAZAR-KLOSINSKI, B., MICHEL, Y., BREWERTON, T. D. & DOBLIN, R. (2013). Durability of improvement in post-traumatic stress disorder symptoms and absence of harmful effects or drug dependency after 3,4-methylenedioxymethamphetamine-assisted psychotherapy: a prospective long-term follow-up study. *J Psychopharmacol,* 27, 28-39.

MITHOEFER, M. C., WAGNER, M. T., MITHOEFER, A. T., MARTIN, S., JEROME, L., MICHEL, Y., YAZAR-KLOSINSKI, B. B., BREWERTON, T. D. & DOBLIN, R. (2012). Safety, Efficacy and Durability of MDMA-Assisted Psychotherapy in Subjects with Chronic, Treatment-Resistant Posttraumatic Stress Disorder: Completed and Ongoing Randomized, Controlled, Triple-blind Phase 2 Pilot Studies. *Military medical research across the continuum of care.* Fort Lauderdale, FL.

NUTT, D. J. & MALIZIA, A. L. (2004). Structural and functional brain changes in posttraumatic stress disorder. *J Clin Psychiatry,* 65 Suppl 1, 11-7.

OEHEN, P., TRABER, R., WIDMER, V. & SCHNYDER, U. (2013). A randomized, controlled pilot study of MDMA (+/- 3,4-Methylenedioxymethamphetamine)-assisted psychotherapy for treatment of resistant, chronic Post-Traumatic Stress Disorder (PTSD). *J Psychopharmacol,* 27, 40-52.

OGDEN, P., MINTON, K. & PAIN, C. (2006). *Trauma and the body : a sensorimotor approach to psychotherapy.* W. W. Norton, New York.

RESICK, P. A. & SCHNICKE, M. K. (1992). Cognitive processing therapy for sexual assault victims. *J Consult Clin Psychol,* 60, 748-56.

SHULGIN, A. & SHULGIN, A. (1991). *Pihkal: A Chemical Love Story.* Transform Press, Berkeley, CA.

STOLAROFF, M. (2004). *The Secret Chief Revealed: Conversations with a pioneer of the underground therapy movement.,* Sarasota FL, Multidisciplinary Association for Psychedelic Studies.

THOMPSON, M. R., CALLAGHAN, P. D., HUNT, G. E., CORNISH, J. L. & MCGREGOR, I. S. (2007). A role for oxytocin and 5-HT(1A) receptors in the prosocial effects of 3,4 methylenedioxymethamphetamine ("ecstasy"). *Neuroscience,* 146, 509-14.

URSANO, R. J., BELL, C., ETH, S., FRIEDMAN, M., NORWOOD, A., PFEFFERBAUM, B., PYNOOS, J. D., ZATZICK, D. F., BENEDEK, D. M., MCINTYRE, J. S., CHARLES, S. C., ALTSHULER, K., COOK, I., CROSS, C. D., MELLMAN, L., MOENCH, L. A., NORQUIST, G., TWEMLOW, S. W., WOODS, S. & YAGER, J. (2004). Practice guideline for the treatment of patients with acute stress disorder and posttraumatic stress disorder. *Am J Psychiatry,* 161, 3-31.

WARDLE, M. C. & DE WIT, H. (2014). MDMA alters emotional processing and facilitates positive social interaction. *Psychopharmacology (Berl).*

THE BIRTH OF MAPS AND MDMA RESEARCH: A PERSONAL REFLECTION

RICK DOBLIN

At the start of its 40-year arch of history, MDMA began as an underground, but not illegal, aid to therapy. After MDMA became more widely used recreationally as Ecstasy, it became criminalised (Lawn, 1986; DEA, 1988). Today the Multidisciplinary Association for Psychedelic Studies, MAPS, is engaged in the effort to bring MDMA back as a prescription medicine.

What began in the mid 1970s was the use of MDMA under the code name *Adam*, gradually spread through Sasha and Anne Shulgin and the leader of their small test group, Leo Zeff, the Secret Chief (Stolaroff 2005). Zeff trained hundreds of psychiatrists and therapists on how to use MDMA. Several of the people who experienced MDMA in that therapeutic context felt that it should be made more widely available, that it could be used in nightclubs for recreational use. In the early 1980s the more widely used MDMA, branded as Ecstasy, could not fail to attract the attention of the police. The police started noticing Ecstasy, mostly in nightclubs in Texas; its popularity was growing (Alexandre, 2009).

I learned about MDMA in 1982. I had no idea that the underground psychedelic networks still existed. It was an incredible relief to find out that, not only did they exist, they also had a legal drug that was quite effective. It was clear, though, that the crackdown was coming. Several other people and I formed a nonprofit called Earth Metabolic Design Lab, a name borrowed from Buckminster Fuller's associates. We decided to organise an effort, in secret, to prepare for the crackdown.

At that time I contacted Robert Muller at the United Nations. In the early 1980s Muller wrote a book called *New Genesis: Shaping a Global Spirituality*; it is very inspiring but he didn't have anything in it about psychedelics (Muller, 1982). I wrote him a letter as a college undergraduate. I said that research, Walter N. Pahnke's Good Friday experiment in 1962, for example, has suggested that psychedelics can help us understand spiritual experience. I asked *would you help us open the door to renew psychedelic research?* To my surprise, he responded. And to my greater surprise, he said yes, he would help.

Muller gave me a list of religious professionals all around the world. The subtext that I read into it was *send them MDMA*, which I then proceeded to do. There were Roman Catholic monks, Buddhist priests, Orthodox rabbis and others who used the MDMA I sent them as a tool, mostly for meditation but also for other ceremonial, ritual purposes. MDMA was very well suited to the spiritual experience. I was hoping these religious professionals would be willing to testify. Once the crackdown came they would be able to say they had used MDMA while it was legal. We also reached out to various psychiatrists and others who were prominent in the field of psychotherapy so we had a base of people prepared who would be courageous enough, and relatively invulnerable enough, to speak.

In 1984 the DEA announced that they wanted to criminalise MDMA (DEA, 1984). We had already done the first safety study of MDMA (Downing, 1986). Once the DEA announced the criminalization (Lawn, 1985) I went to Washington and filed a request for a hearing. The DEA was utterly shocked that I walked in the door because they had no idea about the therapeutic use of MDMA. All they knew was that it was a recreational drug. We won the lawsuit; the judge made a recommendation to the head of the DEA that MDMA should be illegal for everybody *except* for therapists and psychiatrists, for whom it would remain a medicine they could use in their practices (Young, 1986). The DEA ignored that recommendation (DEA, 1986). We appealed again and won, again (Grinspoon, 1986; US Court of Appeals, 1987; DEA, 1988). The DEA came back and rejected the appeal and, eventually, the DEA won (DEA, 1988).

MDMA was off patent; it had been abandoned by the pharmaceutical industry (Merck, 1912). Governments weren't going to pay for it because they were investing in neurotoxicity research to justify prohibition. The major foundations found it too controversial. So, in 1986, I started MAPS with the idea that it would be a nonprofit pharmaceutical company.

It wasn't until 13 years later, in 1999, that the first drug actually became a medicine in a nonprofit context. That was the abortion pill, RU486, funded by the Rockefellers, Warren Buffett, and the Pritzkers, through the Population Council (CDER, 2000). RU486 validated the principle that there could be nonprofit drug development, that it's actually a viable model.

It has taken MAPS 27 years to get to the point where we are now, in the middle of the Phase II research process. In order to develop a drug into a medicine, often it starts with clinical studies in animals and then Phase I safety studies in healthy adults before working in patients, figuring out which specific questions to answer. Phase II is about method (US National Library of Medicine, 2008). For MAPS the method is not just giving MDMA, it's MDMA-*assisted psychotherapy*. We are trying to standardise the psychotherapy aspect. We've developed a treatment manual, available on our website for free (Mithoefer et al., 2013). Independent raters analyse and review sessions to see that the therapists adhere to the methods.

A drug researcher also has to do studies on dose. In our current study, MDMA for PTSD in Charleston, South Carolina, the method includes three doses: 30 mg, 75 mg, and 125 mg (Mithoefer et al., 2012). We've been surprised to find out that many subjects who are getting the 75 mg are getting better. I thought that the 125 mg dose, which is the standard in underground therapy, would be ideal but we are getting remarkable results with 75 mg also. We are honing in on the most effective dosage.

A researcher also needs to identify the patient population. In MAPS' first study we had mostly female survivors of childhood sexual abuse and adult rape and assault. These were people who had PTSD for an average of over 19 years. What we found was that, after treatment, over 80% of them no longer qualified for a diagnosis of PTSD (Mithoefer et al., 2011). Then the question is, what about veterans? Is the same method that worked so well for that group going to work for combat-related PTSD? Our current research leads us to believe that the method works regardless of the cause of PTSD.

Then we get to the biggest problem in this whole area of research, which is the double blind. That's been, historically, one of the main criticisms of the early research in the 1950s and 1960s with psychedelics. At that time psychedelic studies were uncontrolled, there was no effective double blind. There was concern that investigator and subject bias would influence the results. Now there are two basic approaches. One of those is dose response where we give a series of doses and blind subjects to which dose they received. If the lower doses don't work as well as the medium and higher doses then that is an effective double blind. That's the most complicated way because the low dose, in order for it to be confused with the other doses, can't be too low. The more you increase that dose the greater its efficacy. In this case researchers are comparing full dose against something that is already working, an active placebo, making it more challenging to show a difference. It's kind of a secret in clinical research that pharmaceutical companies don't actually try to inquire whether the double blind was effective or not because drugs often have side effects that break the blinding (Mithoefer et al. ,2012; 2013).

These many factors are the core issues of what MAPS is figuring out in Phase II to prepare for Phase III. We anticipate Phase III to cost $15 million. One of MAPS' board members, a brilliant computer person, recently passed away and left us $5 million. We have reserved those funds for our Phase III trials. Our planning shows it will be another couple of years and a couple million dollars to finish Phase II after which MAPS will go forward with design and initiation of Phase III.

For the first time we are operating with all the cylinders firing. By that I mean we have active negotiations going on with the Department of Defense (DoD) and the Veterans Administration (VA). Last year the VA spent $5.5 billion on disability payments to veterans who were unable to work due to posttraumatic stress. (Department of Veterans Affairs 2005) There are massive incentives for research on PTSD. At the same time there is cultural stigma so we have not yet been able to get formal co-operation with the DoD and VA but we are working at higher and higher levels with politicians in Washington.

We also have the good fortune that the current head of the National Institute of Mental Health (NIMH), Dr Tom Insel, did MDMA neurotoxicity research in primates, in the late 1980s (Insel et al., 1989). I went to visit him at that time. I found that he was someone who was doing research, not with the agenda of the National Institute on Drug Abuse (NIDA), which produces science to justify the drug war. Instead, he was doing science for truly understanding the risks, and he was much more moderate in his risk assessment. I contacted him again a couple of years ago and asked, *is now the time for us to apply for an NIMH grant*? Dr Insel said, *yes, it is*. He assigned one of his top PTSD experts to help us craft a proposal to get government funding. What he said was that the peer review committees are notoriously unsympathetic to innovation so he can't guarantee that we can get the funding.

Another recent development in MAPS' planning is the discovery that there is an obscure FDA policy that says that if you are the first to make a medicine, even if all the patents have expired, you can get what's called data exclusivity (FDA, 2014). Data exclusivity means that for a period of five years in the US, ten years in Europe, you can be the only one to market the medicine using that data. Somebody else could use their own studies to get permission to market MDMA but MAPS would be the only one to market MDMA using our data. Data exclusivity would provide for further research.

I estimate that by 2021 MDMA will be a legal prescription medicine. I really believe it's possible.

Alexandre, J. F. (2009). *Warriors of the discotheque: the starck club documentary short version* [Film on DVD]. USA, 2009 June.

Center for Drug Evaluation and Research. (2000). Approval letter addressed to Sandra P. Arnold of Population Council. NDA 20-687. Available: www.accessdata.fda.gov/drugsatfda_docs/appletter/2000/20687appltr.htm

Department of Veterans Affairs. (2005). *Review of state variances in VA disability compensation payments.* Report No. 05-00765-137. Washington, DC: VA Office of Inspector General. 2005 May 19.

Department of Justice, Drug Enforcement Agency. (1984). 21 CFR Part 1308. Schedules of controlled substances; scheduling of 3,4-methylenedioxymethamphetamine (MDMA) into schedule I of the controlled substances act; remand. *Federal register.* Vol. 49. No. 146, 30210. 1984 July 27.

Department of Justice, Drug Enforcement Agency. (1986) 21 CFR Part 1308. Schedules of controlled substances; scheduling of 3,4-methylenedioxymethamphetamine (MDMA) into schedule I of the controlled substances act. *Federal register.* Vol. 51. No. 198, 36552. 1986 Oct 14.

Department of Justice, Drug Enforcement Agency. (1988) 21 CFR Part 1308. Schedules of controlled substances; scheduling of 3,4-methylenedioxymethamphetamine (MDMA) into schedule I of the controlled substances act. *Federal register.* Vol. 53. No. 17, 2225. 1988 Jan 27.

Department of Justice, Drug Enforcement Agency. (1988). Schedules of controlled substances; scheduling of 3,4-methylenedioxymethamphetamine (MDMA) into schedule I of the controlled substances act; remand. *Federal register.* Vol. 53. No. 34, 5156. 1988 Feb 22.

Downing, J. (1986). The psychological and physiological effects of MDMA on normal volunteers. *J Psychoactive Drugs.* Oct-Dec;18(4):335-40.

Grinspoon, L. (1986). Petition for review of Drug Enforcement Administration, Depart of Justice, final order. *United States Court of Appeals for the First Circuit.*

Insel, T. R., Battaglia, G., Johannessen, J. N., Marra, S., De Souza, E. B. (1989) 3,4-Methylenedioxymethamphetamine ("ecstasy") selectively destroys brain serotonin terminals in rhesus monkeys. *J Pharmacol Exp Ther.* Jun;249(3):713-20.

Lawn, J. C. (1985). *Schedules of controlled substances, temporary placement of 3,4-methylenedioxymethamphetamine (MDMA) into schedule I.* Dept. of Just., Drug Enf. Admin. Billing Code 4410-09-M. 21 CFR Part 1308; 1985 May 28.

Merck, E., inventor. (1912). *MDMA.* German patent 274,350. 1912 Dec 24.

Mithoefer, M. C., Mithoefer, A., Wagner, M., Wymer, J. (2012) A randomized, triple-blind, phase 2 pilot study comparing 3 different doses of MDMA in conjunction with manualized psychotherapy in 24 veterans, firefighters, and police officers with chronic, treatment-resistant Posttraumatic Stress Disorder (PTSD). *MAPS Study MP8 Amendment 4 Version 1.* 2012 Feb 6. Available online: http://www.maps.org/research/mdma/MP8_amend4_final_7Feb2012web.pdf

Mithoefer, M. C., Wagner, M. T., Mithoefer, A. T., Jerome, L., Doblin, R. (2011). The safety and efficacy of {+/-}3,4-methylenedioxymethamphetamine-assisted psychotherapy in subjects with chronic, treatment-resistant posttraumatic stress disorder: the first randomized controlled pilot study. *J Psychopharmacol*. 2011 Apr;25(4):439-52. doi: 10.1177/0269881110378371. Epub 2010 Jul 19.

Mithoefer, M. C. (2013). *A manual for MDMA-assisted psychotherapy in the treatment of Posttraumatic Stress Disorder*. Version 6. Santa Cruz, CA: MAPS; 2013 Jan 4. Available from: http://www.maps.org/research/mdma/MDMA-Assisted_Psychotherapy_Treatment_Manual_Version_6_FINAL.pdf

Muller, R. (1982). *New genesis: shaping a global spirituality*. 1st ed. Doubleday.

Pahnke, W. M. (1963). *Drugs and mysticism: an analysis of the relationship between psychedelic drugs and the mystical consciousness*. A thesis presented to the Committee on Higher Degrees in History and Philosophy of Religion. Cambridge (MA); Harvard University; 1963 June. Available online: http://www.maps.org/books/pahnke/walter_pahnke_drugs_and_mysticism.pdf

Stolaroff, M. J. (2005). *The secret chief revealed*. Rev. ed. Sarasota, F. L. MAPS.

US Court of Appeals For the First Circuit. (1987). *Petition for review of order of the Drug Enforcement Administration*. No. 86-2007. 1987 Sep 18.

US Food and Drug Administration [internet]. (2014). *Frequently asked questions on patents and exclusivity*. 2014 Jul 17. Available from: http://www.fda.gov/Drugs/DevelopmentApprovalProcess/ucm079031.htm

US National Library of Medicine [internet]. (2008). *FAQ: Clinical Trials Phases*. National Institute of Health. Apr 18; Available from: http://www.nlm.nih.gov/services/ctphases.html

Young, F. L. (1986). *In the matter of MDMA scheduling: opinion and recommended ruling, findings of fact, conclusions of law and decision of administrative law judge*. Docket No. 84-48. US Department of Justice, Drug Enforcement Agency. 1986 May 22.

BEYOND CASTANADA: A BRIEF HISTORY OF PSYCHEDELICS IN ANTHROPOLOGY

JACK HUNTER

By now the image of the adventurous anthropologist boldly experimenting with the psychoactive substances of their native informants is something of a cliché. Images from Carlos Castaneda's influential series of books, in which a young anthropologist is initiated into the world of Yaqui sorcery through extraordinary psychedelic experiences, immediately spring to mind when the subject comes up. But there is a history of serious anthropological inquiry beyond Castaneda's popularisation (and possible fictionalisation) of anthropology's involvement with psychoactive substances. In this short chapter I aim to give a brief, introductory, chronological summary of developments within this field of study, from the Nineteenth Century to the present day, through presenting snapshots of key figures and their research. These will include, in order of appearance, J. G. Frazer, Weston La Barre, Richard Evans Schultes, Napoleon Chagnon, Carlos Castaneda, Marlene Dobkin de Rios, Michael Harner, Zeljko Jokic and others. It is hoped that this overview will portray the gradual blossoming of a more reflexive, experientially oriented, and above all respectful, approach to studying the cross-cultural use of psychedelic plants and substances.

FROM THE ARMCHAIR TO THE FIELD

Anthropology, like most of the sciences, came into its own as a distinct discipline in the mid-Nineteenth Century. Spurred by the success of Charles Darwin's evolutionary theory in biology, which first emerged in his 1859 book *On the Origin of Species*, many thinkers interested in the study of human society began to construct elaborate evolutionist schemes of social and cultural development: beginning with so-called 'primitive' tribal societies and typically culminating with European society as the most highly developed. For the most part early anthropologists were library-based researchers, fully reliant on the first-hand reports of explorers, adventurers, travel writers and religious missionaries for their research data, including all of their associated cultural assumptions an biases.

Sir James George Frazer was a typical armchair anthropologist. In his epic series of books *The Golden Bough* (1993), a vast collection of traditional rites, rituals, folklore and mythology from around the globe, Frazer refers on several occasions to the use of certain plants for the purpose of producing what he referred to as "temporary inspiration." He describes the prophetess of Apollo's consumption of, and fumigation with, laurel leaves before she prophesied, and explained how the traditional Ugandan priest would smoke a pipe of tobacco "fiercely till he works himself into a frenzy," his loud voice then being recognised as "the voice of the god speaking through him." The widespread use of consciousness altering substances was, therefore, clearly noted by early anthropologists, though their researches rarely went much further than describing (or re-describing), practices observed by missionaries and explorers, barely managing to scratch the surface of particularly complex socio-psycho-cultural phenomena. Indeed, the evolutionist paradigm within which scholars like Frazer were operating essentially blocked any kind of deeper understanding of the role of such substances in different cultures. For Frazer, for example, spirit possession practices involving tobacco, or the use of laurel smoke for inducing prophetic states, were little more than 'primitive' evolutionarily redundant spandrels, already replaced by the ostensibly superior scientific worldview. In other words, beliefs about the efficacy of such substances to put the imbiber in contact with spiritual realities were nothing more than confused interpretations of essentially meaningless experiences of intoxication. From the very outset, therefore, such substances were not *permitted* to have any deeper meaning or value, and were certainly not considered as important or worthy of deeper scholarly attention.

As anthropological thought developed in the early Twentieth Century, participant observation, following the lead of the British-Polish anthropologist Bronisław Malinowski, became anthropology's key tool in the quest to understand culture and society. Malinowski had suggested that the only way to understand a culture was to live in it as completely as possible; to participate in everyday life and, while doing so, to make detailed observations of it (Malinowski, 1922). It was only through this kind of immersive research, so Malinowski argued, that a culture could truly be understood. Writing along similar lines the American ethnographer Franz Boas emphasised the importance of attempting to interpret different cultural systems through their own categories, without the imposition of the ethnographer's own cultural assumptions (Boas, 1920). This was to be a particularly influential idea in Twentieth Century anthropology that would become known as cultural relativism. Naturally, this new emphasis on understanding cultural systems from an insider (or at least near-insider), perspective would have a significant effect on the anthropological understanding of the role of psychoactive plants in different cultures. Unlike the evolutionist paradigm of the Nineteenth Century, with its view of non-Western cultures as primitive and outmoded, the relativist paradigm of the early Twentieth Century would begin to reveal the complexities and inner logics of other cultural systems, which were now understood not as somehow beneath the Euro-American scientific worldview, but as parallel to it.

It wasn't until the 1930s that a concerted effort to investigate the cultural use of psychoactive plants was finally undertaken, with the aim of developing a more complete, and nuanced, understanding than had previously been achieved during the Nineteenth Century. One of the earliest such studies was published in 1938 by the American ethnographer Weston La Barre in his book *The Peyote Cult*, based on his own doctoral research amongst the various tribes of the American plains. In the book La Barre describes the many uses of the peyote cactus (*hikuri*) as a tool for prophecy and divination, as an apotropaic charm of protection to ward off witchcraft and attacks from rival tribes, as well describing its 'technological' use as a medicine for the healing of wounds, curing of snake bites, bruises and many other common afflictions. He even describes the use of the cactus as a cure for blindness.

In addition to these technological uses, La Barre also explored the ritual use of peyote amongst the Huichol and Tarahumara peoples. He describes the traditional pilgrimage of the Huichol to gather peyote as a sacred journey to Wirikuta, the primordial origin of the world, 'since formerly the gods went out to seek peyote and now are met with in the shape of mountains, stones and springs.' When the Huichol pilgrims arrive at the *mesa* where the peyote grow, a ritual is performed in which the peyote cactuses are hunted like deer. The Huichol men fire their arrows over the top of the cactuses, symbolically missing their targets, so that the cactuses may be brought home alive. Rituals, feasts and festivities follow the return of the peyote pilgrims. La Barre emphasised the multidimensional role of the peyote cactus as a central pillar of Huichol culture, in terms of structuring the ritual year, providing access to spiritual realms *and* as a medicinal technology.

In 1940 the ethnobotanist Richard Evans Schultes, a colleague of La Barre, published an account of his research into *teonanácatl*, a hallucinogenic plant reportedly used by the Aztecs. In the Nineteenth Century debates had raged amongst Western scholars concerned with identifying *teonanácatl*, with many assuming that the plant must have been the peyote cactus, owing to its intoxicating effects. But Evans Schultes' reading of the historical documents suggested otherwise, indeed they suggested that *teonanácatl* was in fact a mushroom. However, in order to prove his theory Evans Schultes still needed to identify which particular mushroom *teonanácatl* referred to. And so, with this aim in mind, he embarked on an excursion to conduct fieldwork amongst the Mazatec Indians of the Oaxaca region of Mexico.

In 1938 Evans Schultes tracked down the mushroom *Panaeolus campanulatus* var. *Sphinctrinus* in Huautla de Jiménez, referred to by the Mazatec as *t-hana-sa* (meaning 'unknown'), *she-to* ('pasture mushroom') and *to-shka* ('intoxicating mushroom'). The mushroom grows during the rainy season in boggy spots, and contains small amounts of the psychoactive compound *psilocybin*. Schultes describes how contemporary Mazatec diviners used the mushroom for practical and magical purposes, in order, for example, to locate stolen property, discover secrets and to give advice to those in need. The mushroom was also reportedly used in witchcraft. Schultes describes how consumption of the mushroom induces a "semi-conscious state accompanied by mild delirium" that lasts approximately three hours, over the course of which the subject passes through a period of feeling generally good, a stage of hilarity and incoherence and finally experiences "fantastic visions and brilliant colours," similar in many ways to the peyote experience.

Schultes had found the fabled *teonanácatl* mushroom. Through conducting ethnographic fieldwork amongst contemporary Mazatec Indians, rather than relying solely on the reports of missionaries and explorers, Schultes was able to solve an anthropological puzzle and open the doors for further research on contemporary use of the mushroom amongst the Mazatec.

DEVELOPMENTS IN THE TWENTIETH CENTURY: FROM RELATIVISM TO TRANSPERSONALISM

By the middle years of the Twentieth Century anthropology had blossomed into a fully-fledged academic discipline, and with this blossoming came an increase in the number of detailed ethnographic accounts of the use of psychoactive plants in different cultural contexts.

While pioneering ethnobotanists like Weston La Barre and Richard Evans Schultes continued to investigate the use of psychoactive substances, especially in the context of South America, new experientially oriented approaches were gradually beginning to emerge.

While conducting his doctoral fieldwork amongst the *Jivaro* of eastern Ecuador in the late 1950s, Michael Harner developed a life-long fascination with shamanism, which he would come to recognise as a near-universal set of techniques for experiencing non-ordinary realities. Twice he was offered the chance to partake of the *Jivaro's* psychedelic brew, known as *Natema*, but in an effort to maintain his academic objectivity he declined on both occasions. It wasn't until 1961 that Harner, in his own words, 'crossed the threshold fully' when he drank ayahuasca with the Conibo of eastern Peru (Harner, 2013), an experience that inspired questions about 'the cross-cultural importance of the hallucinogenic experience in shamanism and religion' (Harner, 1973). Harner would go on to conduct a comparative study of themes in Jivaro and Conibo ayahuasca experiences, which revealed common phenomenological tropes including:

- The separation of the soul from the physical body, often associated with sensations of flight and travel
- Visions of jaguars and snakes
- The sense of contact with supernatural beings
- Visions of distant persons, cities and landscapes
- Divinatory visions, for example locating stolen items (Harner, 1973)

Writing in the introduction to his groundbreaking anthology, *Hallucinogens and Shamanism* (1973), Harner looked forward to the influence of ethnographic encounters with psychedelic reality on theory development in anthropology:

> *"...as more anthropologists undertake field research on the significance of hallucinogens and partake of the drugs themselves, it will be interesting to see how 'participant observation' influences their understanding of the cultures studied and affects their personal, theoretical, and methodological orientations"* (Harner, 1973)

Another, by now rather notorious, contributor to this emerging experiential turn was Carlos Castaneda. Castaneda was a graduate student in anthropology at the University of California when he published the famous account of his apprenticeship to the Yaqui sorcerer Don Juan Matus, *The Teachings of Don Juan: A Yaqui Way of Knowledge* (1968).

In the book, and its sequels, Castaneda vividly describes his experience of initiation into the secrets of Yaqui sorcery under Don Juan's tutelage, who taught him how to recognise and prepare several psychoactive plants including peyote (*Lophophora williamsii*) and Jimson Weed (*Datura stramonium*). Castaneda's book revealed the inner workings of Yaqui sorcery, exploring the experiential psychedelic underpinnings of Yaqui supernatural belief. Although there has been a great deal of debate regarding the veracity of Castaneda's account (see Murray, 1979), his books have had an undoubted influence on many ethnographers investigating both traditional belief systems and psychoactive plant use.

In 1971 the similarly controversial (though for entirely different reasons) American ethnographer Napoleon Chagnon and his colleagues published a paper entitled *Yanomamo Hallucinogens: Anthropological, Botanical, and Chemical Findings*. The Yanomamo people inhabit small villages in the Amazon rainforest of Venezuela and northern Brazil, and, amongst other things, are well known for their use of the psychoactive snuffs *epene* (a general term for any prepared snuff), and *yopo* (*Anadenanthera peregrina*). Chagnon and his colleagues, as part of their more traditional ethnobotanical research, were able to get hold of samples of the *epene* snuff to be subjected to chemical analysis in the laboratory. In addition to various non-psychoactive ingredients, the *epene* snuff was found to contain high levels of the psychoactive 5-hydroxy-N,N-dimethytryptamine, or Bufotenine, very closely related to the somewhat more famous compound known as DMT – see Strassman (2001) for more on this unusual compound. These hallucinogenic snuffs, as we shall soon see through the recent work of Zeljko Jokic, constitute an integral component of Yanomamo culture and cosmological understanding.

In the 1970s American anthropologist Marlene Dobkin de Rios carried out extensive research into the cultural use of psychoactive substances in several contemporary and historical South American societies. Her research included examining the influence of hallucinatory experience on Maya, Mochica and Nazca religion and art, as well as examining contemporary uses of psychoactive substances, including ayahuasca, for folk healing rituals and witchcraft in both rural and urban settings. In a 1975 paper prepared for the *National Commission on Marihuana and Drug Abuse,* which draws on case studies of societies in which psychoactive plants play a significant cultural role, de Rios highlighted new potentials for drug use in Western societies. Her sample of case study societies included Siberian Reindeer herders, and their use of the Fly Agaric mushroom (*Amanita muscaria*), the Shagana-Tsonga of the Transvaal and their ritual use of *Datura fastuosa*, the Amazonian Amahuaca's use of *yage*, and the Peruvian uses of the San Pedro cactus (*Trichocereus pachanoi*). From these disparate examples de Rios highlighted several key factors in the traditional ritual use of psychoactive substances, which she categorised as biological (including body weight, special diets, sexual abstinence, etc.), psychological (mindset, personality, mood, etc.), social-interactional (nature of the group, ritual performance, presence of a guide, etc.) and cultural (including a shared symbolic system, belief system and so on). De Rios argued that through paying attention to these significant variables a more meaningful interpretation of drug experiences, as well as a more sensible attitude towards drug consumption, might be achieved in the context of Western drug use. As anthropologist and ethnobotanist Wade Davis has recently put it:

> "...whatever the ostensible purpose of the hallucinogenic journey, the Amerindian imbibes his plants in a highly structured manner that places a ritualistic framework of order around their use. Moreover the experience is explicitly sought for positive ends. It is not a means of escaping from an uncertain existence; rather it is perceived as a means of contributing to the welfare of all one's people." (Davis, 2010)

This has, I believe, been a particularly important realisation for anthropology, as well as for Western attitudes towards drug consumption. Experiences once seen as essentially meaningless, and understood through the Euro-centric lens of 'intoxication,' are finally being recognised as meaningful, both individually (on the personal scale), and socio-culturally. Transpersonal anthropologist Charles Laughlin, for example, has characterised different cultural attitudes to altered states of consciousness through his classification of monophasic and polyphasic societies. He writes:

"Many societies integrate knowledge gleaned in experiences in all phases of consciousness within a single worldview. We call these polyphasic cultures. By contrast, our own society typically gives credence only to experiences had in the 'normal' waking phases – that is, in the phases of consciousness oriented primarily toward adapting to the external operational environment. We thus live in a relatively monophasic culture."   *(Laughlin, 1992)*

This move towards an understanding of altered states of consciousness as functional, desirable, and above all meaningful to those who experience them, has opened up new avenues for anthropological investigation.

INTO THE TWENTY-FIRST CENTURY

An interesting development in recent anthropological research has been the move towards investigating the use of psychedelics in 'Western' societies. Tramacchi (2000), for example, has conducted fieldwork at psychedelic gatherings known as *doofs* in the Australian bush, which emerged as a response to the perceived commercialisation of raves, coupled with a desire to ritualise psychedelic use for spiritual means. Tramacchi explores Victor Turner's concept of *communitas*, a sense of common experience and communal bonding, in the context of the *doof*:

"Collective rituals which incorporate potent psycho-active sacraments can stimulate profound subjective individual experiences, but they are simultaneously a socially dynamic collective force."

(Tramacchi, 2000)

Again, Tramacchi recognises the role played by psychedelics, in conjunction with what he terms 'ecstatic ingredients' (the liminal quality of camping, psychedelic music, kaleidoscopic light shows, religious iconography, dancing, etc.), as both individually and communally meaningful. Explorations of psychedelic use and experience in 'Western' societies represent an exciting new arena for ethnographic investigation (cf. Adams, 2012; Gelfer, 2012).

More recently, ethnographer Zeljko Jokic, in his 2008 paper on the initiation of Yanomamo *shapori* (shamans), gained significant insight into the experiential life-world of the people he was studying. Through participating fully in the *shapori* initiation ceremony, including the ingestion of the psychoactive snuff *yopo* (*Anadenanthera peregrina*), Jokic was able to experience culturally significant states of consciousness that simply could not have been encountered in any other way. For instance, Jokic describes his experience looking up the *pei maki* ceremonial pole, lodged between his legs:

> *Experientially, there was no sense of separation of space between my ego and the pole. The two (subject – object) merged into a single dimension of consciousness while simultaneously a new sense of space emerged. As I was observing the pole, it appeared huge, stretching infinitely from one extreme of the cosmos to the other. Although my consciousness was altered and my ego fully immersed in the experience, simultaneously a part of me was still engaged in reflexive thinking. At that moment, I thought to myself: "This must be the* axis mundi *which connects all spheres of the universe."*

Experiences of being dismembered by spirits, transforming into different forest creatures, such as snakes and jaguars and assuming their perspectives, also accorded well with Yanomamo beliefs and cosmology. This kind of immersive participation leads to a new appreciation of the cosmological significance of experiences with psychedelics, and leads to intriguing possibilities in exploring the experiential source hypothesis (Hufford, 1982), as it relates to systems of 'ethno-metaphysics' (Hallowell, 2002).

Jokic explains how:

> *"I treat my own experiences not only as subjective, isolated events of my own intentionality of consciousness but as an intersubjective dialogic product arising within the interpersonal field of social relations. Shamanism for me is a point of intersubjective entry into the Yanomami lifeworld, and the resulting experiences are the window that provides my own subjective insights into that world."*
>
> *(Jokic, 2008)*

Through participating in the Yanomamo shamanic initiation, and experiencing culturally significant states of consciousness, Jokic was able to 'become one' with his hosts, to experience the world as they do, at least for the duration of his experience.

CONCLUSIONS

Andy Letcher recognises two dominant discourses in academic approaches to the study of psychedelic mushrooms, which can be equally applied to other psychedelics. The first set, which he defines as 'pathological,' 'psychological,' and 'prohibition' discourses, derives from objective observations of the effects, the symptomatology, of psychoactive substances on others. The second set of discourses, including 'recreational,' 'psychedelic,' 'entheogenic' and 'animistic,' emerges in opposition to these discourses and derives from practitioners themselves (Letcher, 2007). What we have seen in this brief history, then, is an expression of the friction between these dominant sets of approaches, and of the deficiencies of the discourses that have dominated academic research in this area, progressing from the pathological and prohibitionist discourses of the Nineteenth Century to a more reflexive, experiential, and sensitive understanding.

Although not a complete history, a trend towards a more reflexive and experiential approach has emerged. Anthropological approaches to the use of psychoactive plants have gradually changed with the development of the discipline's underlying paradigm. Nineteenth Century approaches were limited by the assumptions of the evolutionist paradigm, according to which non-Western cultures were seen as somehow 'primitive,' 'superstitious' and already superseded by the Euro-American scientific worldview. By the beginning of the Twentieth Century this assumption was being questioned, with non-Western cultures beginning to be seen as parallel with, rather than subordinate to, Western culture. The cultural relativist paradigm would lay the foundations for further developments in anthropological approaches to the study of psychoactive plants. By the middle of the Twentieth Century a new experiential approach emerged, placing an emphasis on the experiential foundations of traditional belief systems and paving the way for new 'transpersonal' modes of understanding the value of psychedelic experiences in both Western and Non-Western societies.

Allen, M. R. (1967). *Male Cults and Secret Initiations in Melanesia.* Melbourne University Press, Melbourne.

Bowers, N. (1965). Permanent Bachelorhood in the Upper Kaugel Valley of Highland New Guinea. *Oceania* 36,1: 27–37.

Dosedla, H. (1974). Ethnobotanische Grundlagen der materiellen Kultur der Mount Hagen-Stämme im zentralen Hochland von Neuguinea. Tribus, *Jahrbuch des Linden-Museums für Völkerkunde Stuttgart,* Vol. 23, Stutttgart 1974

Dosedla, H. (1978). Oral Tradition, Historical Consciousness and Cultural Change among the Mbowamb of the Central Highlands of Papua New Guinea. *Wiener ethnohistorische Blätter* 16: 91–112.

Dosedla, H. (1984). Fishing in the central highlands of Papua New Guinea. Pages 1115-1143. *The Fishing Culture of the World: Studies in Ethnology, Cultural Ecology, and Folklore.* Vol.2, University of London

Dosedla, H. (2011). Mushroom and Masalai Madness in Melanesia - Drug traditions and cultural change in highland societies of Papua-New. *Paranthropology: Journal of Anthropological Approaches to the Paranormal.* Vol. 2 No. 4 (PP 36-41).

Dosedla, H (2012). The Kind and Unkind Papuan Girls - Tumbuna Tales from the Highlands of Papua New Guinea. *Fabula* VOL: 53; 1-18.

Dosedla, H. (2014). *Between animism, demonism and divine hierarchism - Interrelations of mythological concepts between tribal societies of New Guinea and the Philippines.* Paper presented at the Inaugural National Conference of Philippine Association for the Study of Culture, History and Religion (PASCHR). Hosted and co-organized by Holy Angel University Angeles City, Pampanga, Philippines, January 31 to February 1, 2014

Holden, J. M. (2009). Veridical perception in near-death experiences. Holden, J. M., Greyson, B., & James, D. (Eds.), *The Handbook of Near-Death Experiences: Thirty years of* Handbook of Near-Death Experiences. Library of Congress Cataloguing in Publishing Data. 185–211. Praeger Publishers, Santa Barbara, CA.

Strathern, A. and M. (1971). *Self-decoration in Mount Hagen.* London 1971.

Strauss, H. & Tischner, H. (1962). Die Mi-Kultur der Hagenbergstämme im östlichen Zentral-Neuguinea. *Monographien zur Völkerkunde III.* Chapter 54, PP 403-436. De Gruyter, Hamburg.

Vicedom, G. F. & Tischner, H. (1943-48). Die Mbowamb. Die Kultur der Hagenberg-Stämme im östlichen Zentral-Neuguinea 1–3. *Monographien zur Völkerkunde I.* Hamburg.

"SUPERNATURAL ENCOUNTERS" – ANTHROPOLOGICAL CASE STUDIES FROM INTERIOR NEW GUINEA

HENRY DOSEDLA

ABSTRACT

Having dealt with significant psychedelic states marked by changes of awareness induced by the consumption of distinct mushrooms and other herbal drugs in another publication (Dosedla, 2011), the topic of this paper is focused on reports on psychedelic experiences based on other causes. These range from hallucinations and synesthesia to trance, hypnotic or mystical states. These may lead to changes in mental operations that are different from normal states including confusion and psychosis that eventually result in the perception of various supernatural encounters. Apart from countless spiritual beings playing an important role within traditional oral lore and folk beliefs, there are also various reports of supposed actual encounters with such entities by members of tribal societies from most parts of New Guinea. In the course of my anthropological fieldwork in the interior highlands there during the early 1970s, when the local population just had experienced primary contacts with modern civilisation, I became familiar not only with many indigenous informants claiming to have had strange manifestations of that kind, but also had chances to witness such occasions myself. The aim of this paper is to give accurate descriptions of these events as well as an approach towards a dispassionate interpretation according to adequate anthropological methods.

INTRODUCTION

According to anthropological records most highland tribes of Papua New Guinea (PNG) share similar concepts of a spiritual world with the inhabitants of the coastal regions and the surrounding smaller islands, which include a similar set of several types of spiritual beings. Though soon after primary European contact missionaries of all confessional shadings started a random competition towards baptising as many 'pagans' as possible, this never resulted in termination of traditional spirit beliefs. On the contrary, most converts since then hesitate to engage in worshipping spirits but more would fear and believe in them.

While PNG highlanders never worshipped any deities they still believe in some kind of supreme spirits, referred to as "sky people" who figure in ancestral myths but are considered to never contact humans.

The greater part of the spiritual universe is divided into ancestral spirits and other kinds of spirits. Regarding human encounters with ancestral spirits, the only reports deal with their occasional invisible appearance at night or, more frequently, their occurrence in dreams. Other human contacts with supposed members of the spiritual world happen with several types of spiritual beings, which are dealt with in detail in this essay. Reported encounters with such unusual manifestations apparently follow distinct patterns, which will be described likewise. Also discussed are the analyses of distinct occasions of such encounters, of the opinions of persons engaged in them and the probable state of minds they had during their extraordinary experiences.

TYPOLOGICAL CHARACTERS OF THE REGIONAL SPIRIT WORLD

TRADITIONAL CHARACTERS

The spirit world of the highlanders is generally classified into mountain spirits, water spirits and several kinds of bush spirits. While mountain spirits, considered to belong to the legendary "sky people", are solely figuring in ancestral myths, there is one case from the Mount Hagen tribes of the Western Highlands Province which is reported to have happened shortly before the arrival of the first missionaries in the region, when it was taken as an evil omen. According to eye witnesses of that event, a gigantic dark leg coming down from the clouds stood above the ground for several hours during an unusual darkness, which most likely could be interpreted as a kind of tornado, twister or 'land spout'.

There are other reported encounters with rock spirits and water spirits by persons who had happened to become lost or buried in caves by accidents, possibly also having lost consciousness, after which they claimed actually having seen such supernatural beings then. The same applies to encounters with water spirits which follow identical patterns, since they were reported by persons who nearly had become drowned in lakes, pools or rivers, especially subterranean water streams, considered as the realms of the underworld (Strauss & Tischner, 1962).

Significantly descriptions of such experiences apparently induced by states of extreme shock exactly match with typical features known from traditional narratives, which begs the question whether it was the hen or the egg having appeared first. Considering such episodes as cases of near-death experiences (NDE) this could explain the cause of certain hallucinations in close connection to the cultural beliefs held by the individual, which seem to form the phenomena witnessed in the NDE and the later interpretation thereof (Holden, 2009).

Regarding water spirits as a significant tradition in the Western Highlands, there is a widespread cult based on the mythical narration of a man who on passing a subterranean water stream met such a demon whom revealed to him the secrets of a distinct fertility cult. Cult members used to re-enact that story in a series of intricate acts, including the eventual consumption of a drug of the *Euphorbia* family, known as effective in poison fishing (Dosedla, 1984). This causes states of stupefaction and disorientation as well as claustrophobia and the impression of passing through narrow tunnels, as in cases of a near-death experience and may also be accompanied by acoustic hallucinations (Dosedla, 2011).

According to narrative lore that water spirit, known by various names as "Kör Wöp", "Eimp", "Timp(u)" or "Rimbu", is visualised as an ugly old woman with an enormous protruding mouth and rugged backside – like a toad or even a crocodile – though the latter are completely unknown in interior New Guinea (Dosedla, 2012) and are represented by wickerwork figures and prehistoric stone sculptures.

Another spirit cult of the Western Highlands known by the name of "Kör Nganap" is visualised as a beautiful young woman with a bright face and a crown like headdress consisting of a cloud of shining cobwebs. She is imagined as a virgin, of the same kind as those of the mythical "sky people", and her worshippers would expect to gain more masculine power and protection from pollution by menstruating women (Strathern, 1971).

Eventual encounters with this spirit, which are significantly reported by bachelors suffering from problems caused by sexual deprivation, therefore are considered as an extremely benevolent omen.

Apart from spirits getting worshipped there are other kinds that are just feared. In the Western Highlands these are distinguished by different vernacular names. While the term "Kör" is used for individual spirits imagined to bear some distinct shape the term "Kom" is reserved for rather shapeless spirits (Strauss & Tischner, 1962). Since the gradual replacement of these terms by the common expression "Masalai", indicating some evil bush boogie deriving from the coastal regions, both types have become rather mixed up.

Most reported cases are encounters with a type known as "Ndepona Kör" (forest spirit) described as bearing long protruding tusks and claws who is feared for abducting human victims into the wilderness and raping women. Another type of the same wicked reputation known by the name

"Pim" is described as extremely hairy and bearing his mouth on top of his skull (Dosedla, 2014).

The spirit type known as "Kom" is feared for his ability to incubate his victims; resulting in their rapidly increasing insatiable appetite. This belief was the cause of the former custom regarding the birth of twins, since one of the babies was considered to be a spirit child and was subsequently usually killed. Another typical feature of a "Kom" spirit is his reputed ability to act as a "changeling" in the case of babies, or a "look alike" in the case of adult victims. To my personal experience I met several local people who insisted they had seen my "look alike" at a distant place when actually I was never there. To my surprise the same was even reported to me by friends in Europe, who claimed to have seen my "look alike" there when I was actually overseas (Dosedla, 1914).

NON-IDENTIFIED CHARACTERS

There is an apparent difference between supernatural encounters that members of a local group experience with types of spirits of their own familiar tradition, compared to unexpected encounters with other types of spirits belonging to different traditions unknown to them. A number of such cases happened within a compound of contract labourers from the Southern Highlands Province who lived on plantations on the vast Wahgi Plain in the Western Highlands Province, where they were an isolated group of Kewa tribesmen surrounded by superior numbers of potentially hostile Mount Hagen tribes. This may explain their permanent state of considerable stress making them capable of a set of unpleasant experiences ranging from severe headaches, heart troubles and panic attacks to all sorts of mental disorders.

Whilst there during my research period between 1971 and 1972 I happened to witness a series of apparent mass spirit encounter phenomena, all following the same distinct patterns. Usually plantation workers spent the time in their simple bamboo huts designed for six to eight persons; sharing fire place yarns or enjoying gambling. On one such occasion one of them jumped suddenly up, screaming with fear and running out into the wilderness as if being haunted. In some cases the affected person in this state of panic would push his arm through a gap in the wall of the hut made of plaited bamboo strips with such cruel force that his skin was severely ripped off, while shouting he was getting pulled outside.

Even with the combined strength of his comrades it would be hard work to clutch him firmly enough and pull him back and he would inevitably escape from them and run off into the bush. In most cases it took several days until victims of such attacks were found or brought back from far away; still in a state of trembling, stumbling and unable to explain what had happened to them. Since neither his Kewa mates nor local Mount Hagen tribesmen could tell what type of spirit might be responsible for such behaviour the most common explanation is that it was the work of some unknown "Masalai" or bush boogie (Dosedla, 2011).

Another unidentified spirit type was a frequent manifestation at the same place within the swampy Wahgi Plain, which I also happened to witness several times. Its appearance was marked on full moon nights by some insistent scratching at the wattle of houses accompanied by marked gaps within the bamboo texture, causing the impression that some medium sized invisible creature had been climbing up and down or along the walls. Since this was such a common experience it was evident that it could not be caused by any animal. Everyone there shared the opinion that this was a ghostly but rather harmless woman. Some persons claiming to have seen her at times walking around with a digging stick and supposed that she might be in search of ancestral bones, which frequently used to be found in the prevailing swamp grounds.

A similar case of some invisible creature also happened once during a full moon night at the same place. There was stamping at low pace along the large tin roof of a plantation building, which pressed down by every step as if under considerable weight. Awakened by the noise caused by these gigantic steps all the workers were standing around watching until I took a ladder, and in spite of warnings that I could get thrown down, went on top myself to have a closer look. According to local tradition, as soon as I had addressed the invisible manifestation in a loud voice asking for the purpose of the visit, the noisy steps stopped and never came back.

Some spirits may appear in animal shape as in the case of pigs or dogs, but such reported events are significantly rare. An exception is the demoniac figure of a ghostly tree-climbing kangaroo, common in the Southern Highlands. While "sikau" is the common term for tree kangaroo there is a belief in a werewolf – an animal of man size, which would attack hunters not only with teeth, claws and strokes of his tail but also by weapons such as a spear, axe or bush knife (Dosedla, 1914).

VISIONARY EXPERIENCES

There are many reported cases of all sorts of animals ranging from birds, fish or rats to insects, as well as distinct plants, which act as a kind of messenger between humans and the supernatural world. Distinct occasions of such encounters known by the vernacular "ugl" play an important part within ancestral mythologies in the Western Highlands (Vicedom & Tischner, 1943).

A spectacular case of mass hysteria happened on August 15th 1971 at the main square of the provincial capital Mount Hagen where hundreds of customers from all around had gathered as usual on a market day. Around noon suddenly a mass panic arose from rumours that a supernatural voice had been heard. Many of the irritated people furiously claimed they had witnessed a lamentation emerging from a heap of sweet potatoes in one corner of the market. According to them the tubers were crying like babies, complaining they had been left uncovered in spite of the burning sun and were thus in danger of drying out, getting wrinkled and rotting. Soon the woman in charge of the potatoes was found at a nearby store house where she had gossiped with a couple of friends from a neighbouring village while her goods had remained unprotected. The crowd was shouting at her and near to beating her up because the event was taken as an evil omen that some poor harvests and probable starvation were to be expected as a punishment for such a careless and impious treatment of her precious foodstuff.

According to regional belief no one is expected to sleep outside a house during the night or daytime. In exceptional cases – as on a hunting or trade party for several days – there is no sleeping without previously taking intricate magical precautions. This is explained by the capability of a sleeping person to fall victim to evil spiritual influences. Still it often happens that a person might eventually fall asleep while resting briefly during the daytime. If this were noticed by a companion they would carefully watch the sleeping person in order to see from his or her reactions whether they might appear spiritually endangered. As soon as the person wakes up he or she is eagerly asked if there were any daydreams, since these are considered as an important visionary state (Strauss & Tischner 1962).

For similar reasons there is a common habit of all members of a household gathering together and discussing their dreams first thing after awaking in the morning. In the case of a dream being considered a bad omen the person believed to be endangered is suggested to stay at home that day in order to avoid any bad accidents. Apart from ordinary persons sharing such abilities, there are also individuals in every tribal community reputed as so-called ritual experts acting in a shamanic way by intentionally inducing states of trance in order to perceive revealing visions (Strathern, 1971; Dosedla, 2011). As a matter of fact, among the numerous reported cases of such of predictions at the time of my field studies, I witnessed some coming true myself.

CONCLUSION

Many of the distinct features of mythological highlands lore are not only shared by most other traditional societies of New Guinea, but to a considerable extent are also seen among those of the neighbouring archipelago of the Philippines. These striking similarities could be explained by completely separate origins but may also be associatedwith probable maritime links between both archipelagos that could be traced back into still unknown prehistoric periods.

Adams, C. (2012). 'Psychedelics, Spirits and the Sacred Feminine: Communion as Cultural Critique.' *Paranthropology*, Vol. 2, No. 3, 49-52.

Boas, F. (1920). 'The Methods of Ethnology.' *American Anthropologist*, Vol. 22, No. 4, 311-321

Castaneda, C. (1976). *The Teachings of Don Juan: A Yaqui Way of Knowledge*. Penguin Books, Harmondsworth.

Chagnon, N., Le Quesne, P. & Cook, J.M. (1971). Yanomamo Hallucinogens: Anthropological, Botanical and Chemical Findings. *Current Anthropology*, Vol. 12, No. 1, 72-74.

Davis, W. (2010). *Hallucinogenic Plants and Their Use in Traditional Societies: An Overview*. Available at: http://www.culturalsurvival.org/publications/cultural-survival-quarterly/botswana/hallucinogenic-plants-and-their-use-traditional-so. 13/12/12

Dobkin de Rios, M. (1975). Man, Culture and Hallucinogens: An Overview. In Vera Rubin (ed.) (1975) *Cannabis and Culture*. 401-417. Aldine Publishing Company, Chicago.

Frazer, J. G. (1993). *The Golden Bough: A Study of Magic and Religion*. Wordsworth Editions, London.

Gelfer, J. (2012). Entheogenic Spirituality and Gender in Australia. *Paranthropology*, Vol. 3, No. 3, 22-33.

Hallowell, A. I. (2002 [1960]) 'Ojibwa Ontology, Behaviour, and World View.' In Harvey, G. (ed.) (2002). *Readings in Indigenous Religions*. Continuum, London.

Harner, M. J, (1973). *Hallucinogens and Shamanism*. Oxford University Press, Oxford.

Harner, M. (2013). *Cave & Cosmos: Shamanic Encounters With Another Reality*. North Atlantic Books, Berkeley.

Hufford, D. J. (1982). *The Terror That Comes in the Night: An Experience-Centred Study of Supernatural Assault Traditions*. University of Pennsylvania Press , Philadelphia.

Jokic, Z. (2008). Yanomami Shamanic Initiation: The Meaning of Death and Postmortem Consciousness in Transition. *Anthropology of Consciousness*, Vol. 19, No. 1, 33-59.

La Barre, R. W (1969). *The Peyote Cult*. Schocken Books, New York.

Laughlin, C. D. (1992). Consciousness in Biogenetic Structural Theory. *Anthropology of Consciousness*, Vol. 3, No. 1-2, 17-22.

Letcher, A. (2007). Mad Thoughts on Mushrooms: Discourse and Power in the Study of Psychedelic Consciousness. *Anthropology of Consciousness*, Vol. 18, No. 2, 74-97.

Malinowski, B. (1922). *Argonauts of the Western Pacific*.

Murray, S. O. (1979). The Scientific Reception of Castaneda. *Contemporary Sociology*, Vol. 8, 189-196.

Schultes, R. E. (1940). "Teonanacatl: The Narcotic Mushroom of the Aztecs." *American Anthropologist*, 42.3, 429-443.

Strassman, R. (2001). *DMT: The Spirit Molecule*. Park Street Press, Rochester.

Tramacchi, D. (2000). Field Tripping: Psychedelic Communitas and Ritual in the Australian Bush. *Journal of Contemporary Religion*, Vol. 15, No. 2, 201-213.

SOMA & AYAHUASCA

MATTHEW CLARK

INTRODUCTION

The identity of the plant known as *soma* in ancient India, and as *haoma* in the Zoroastrian tradition, has for around 250 years exercised the wits and imaginations of dozens of scholars. This plant is praised in the highest terms – as a kind of deity – in both Vedic and Zoroastrian texts that date from around 1500 BCE; it is said to provide health, power, wisdom and even immortality.

The four *Veda*s (*Ṛgveda, Yajurveda, Sāmaveda, Atharvaveda*) are a corpus of primarily oral, religious texts composed between about 1500 and 800 BCE, containing mantras and hymns that are still ritually recited (those from the *Sāmaveda* are chanted) in many brahman families both domestically and occasionally in 'public' rituals in India known as *yajña* (praise/worship/sacrifice).

The teachings of the prophet Zoroaster are contained in the *Avesta* (which comprises 72 chapters/sections), the earlier portion of which comprises 17 *gatha*s, which date from perhaps around 1200 BCE. *Gatha*s are recited in daily Zoroastrian worship, known as *yasna*. Some of the *yasna* comprises *yashts*, which are hymns to semi-divine beings, including Hōm (i.e. *haoma*).

The *soma/haoma* plant that is described and praised in these texts has been variously identified by researchers as a non-psychoactive plant, as a medicine, as alcoholic, as a 'narcotic', as a stimulant, and as a psychedelic. Over 50 theories have been produced. Two of the most comprehensive summaries of the published theories to date are those of O' Flaherty (1969) and Houben (2003). The thesis being presented in this article is that *soma/haoma* was in all probability not a single plant but a combination of psychoactive plants, and that in ancient Asia in the late Bronze Age (c. 1500 BCE) there was sufficient botanical knowledge to produce an entheogenic concoction of 'anahuasca' (an analogue of ayahuasca). Although definitive proof of which plants were used has not been established, there are several important clues in the texts consulted that indicate future lines of inquiry for chemical analysis of potentially psychoactive properties of particular plants.

SOMA IN THE VEDAS

In the *Veda*s, after Indra and Agni, Soma is the third-most mentioned deity. Besides being a deity, *soma* has two other aspects: as a plant and as the juice of a plant. According to the *Vedas, Brāhmaṇas, Śrautasūtra*s and other commentarial texts, the *soma*, in the form of bundles of stalks (of a plant with shoots but no leaves), is usually purchased from a *śūdra* (=low-class), in exchange for a cow (and sometimes other goods in the form of gold or a goat); the deal is haggled. The best quality is said to come from the mountain Mūjavat. The bundles of *soma* stalks are examined and extraneous plants are weeded out. With the recitation of mantras, the stalks are then sprinkled with warm water to make them swell, after which they placed on a bull's hide and handfuls of stalks are pounded with stones on two planks of wood to extract the juice, a process undertaken inside the Vedic ritual arena. Sometimes *soma* is pressed with mortar and pestle (*Ṛgveda* 1.28.1–4). Pressings of *soma* (from the root √*su*/√*hu*, meaning 'press') are performed three times a day, in the morning, noon and evening. The extracted juice, which is said to make a lot of noise, like a bellowing bull, is passed through a sheep-wool filter, from which it issues 'flowing clearly'/'purified' (*pavamāna*). It is then mixed with specially drawn water, which has been left standing overnight, in wooden casks. Milk and usually also curd and barley are then added to the water and the extracted *soma* juice. The mixture is offered to the gods on a litter of grass that has been carried on the *soma* cart to the ritual enclosure and is then drunk from bowls/cups by the priests who officiate at the ritual.

Soma is the 'king' of plants; its effects are compared to (the strength) of a bull; it is swift like a steed, brilliant like the sun, conferring immortality on gods and men; it has medicinal power and confers long life; it destroys falsehood and promotes truth; it stimulates the voice and inspires poets and the composing of hymns; it is all knowing. *Soma* is several times said to grow on mountains, from where (or from heaven) it was brought by a falcon (*śyena*). Adjectives used for its colour, which are not always easy to determine, include *hari* (green/yellow/golden), *babhru* (brown/tawny), and *aruṇa* (red/brown/tawny). It is occasionally called 'sweet' (*madhu*), but that seems to be only after the addition of milk, which makes it milder; otherwise it seems to be 'sharp' (*tīvra*) or bitter/piquant/astringent. Strictly speaking, '*soma*' in the *Veda*s is not a proper name but is a ritual designation for a substance out of which *soma* was pressed (Elizarenkova, 1996; Hillerbrandt, 1980).

HAOMA IN THE AVESTA

In the *Avesta*, chapters/sections 9–11 (the *Hōm Yasht*) are litanies to *haoma* (*hōm*), which, as in the *Veda*s, is both a plant and its juice, and a form of deity/spirit being (*yazad*) (Josephson, 1997). When properly prepared and praised, *haoma* bestows six gifts: heaven, health, long life, power to prevent evil, victory against enemies, and fore-warnings against thieves and murderers (ibid.). It also helps barren women desiring children (ibid.) and maidens to find husbands (ibid.). It grants knowledge and wisdom to those who have long sat, searching in books (ibid.). *Haoma* is described as being green and also as yellow (ibid.) or golden-coloured, with pliant shoots (ibid.), giving power and health to the whole body (ibid.) and ecstasy (ibid.). It appears to have branches and sprigs (Modi, 1922). *Haoma* is said to be "of many kinds" (*Hōm Yasht* in Josephson, 1997).

Parpola (1997) maintains that the cult of drinking *haoma* was evidently adopted by Zoroastrianism from the earlier Bronze Age religion of Central Asia and eastern Iran. In contemporary Zoroastrian practice, in both India and Iran, it is the ephedra plant, which can be used as a mild stimulant, that has been used as *haoma* for many centuries. Although there are alternative theories to that of ephedra (see below), most scholars of Zoroastrianism (see Modi, 1922; Boyce, 1975; Gnoli, 2005) are of the opinion that ephedra was the 'original' *haoma*.

WHAT ARE THE EFFECTS OF SOMA/HAOMA?

The effects of *soma* are frequently mentioned in the *Veda*s and also in the *Avesta* (as *haoma/hōm*), as not only exhilarating but as intoxicating (*mada*). This term is used 279 times in the *Ṛgveda*; around 400 times if other compound and variant forms like *madira* are included (Thompson, 2003). This term may be used for someone under the influence of alcohol, but contrasts are occasionally provided in the *Veda*s and *Brāhmaṇa*s between the 'base' intoxication from alcohol (*surā*) (see Brough, 1971) and the 'elevated' intoxication of *soma*. A similar distinction is made in the *Avesta*, between the 'fury' that all other intoxicants produce, and 'gladdening Truth', which only *haoma* reveals (Gershevitch, 1974). The term *mada* is difficult to translate adequately; Brough (1971) suggests something like "possession by a divinity", which stimulates poetic creativity; Staal (2001) notes its range of meanings, including delight, intoxication, inspiration, rapture and elation. Boyce (1975) comments that the drink:

> *"exhilarated and gave heightened powers; and this was the only intoxicant (madha) which produced no harmful effects...the madha of haoma is accompanied by its own rightfulness (aśa)."*

However, it seems that it was possible to drink too much *soma*, as there are *mantra*s in the *Yajurveda* (19–21) for the *sautrāmaṇi*, a ceremony recommended to expiate and counteract the effects of excessive *soma* drinking (Gonda, 1975). This ceremony is recommended for someone who has either drunk too much *soma* or who cannot endure it.

In the *Labasūkta* (*Ṛgveda* 10.119: 'song of the lapwing') the poet describes how he has been "inspired" and swiftly "lifted up" after drinking *soma*, as though by raging winds, impelled upwards to the clouds. Thompson (2003) maintains that this passage, being an *ātmastuti*, is describing the poet's own experience, whatever the cause. Sturhman (1985; 2006) similarly emphasises the visionary aspect of the poet's experiences. Another oft-cited verse of the *Ṛgveda* 8.48.3–4 states:

> *"We have drunk the soma, we have become immortal, we have attained the light, we have found the gods."*   *(trans. Gonda, 1975)*

In *Ṛgveda* 9.113, the poet expresses the desire to reach, through *soma*, the immortal and imperishable world, in eternal light, in the interior of heaven. Commenting on *Ṛgveda* 9.107.20, Elizarenkova (1996) maintains that *soma* clearly causes visions. Numerous times *soma* is said to produce *vipra* (inspiration/trembling). There are also indications that *soma* could cause vomiting; it is also a purgative (Gonda, 1982).

Regarding the effects of *haoma*, Flattery and Schwartz (1989) observe that in ancient Iranian religion there are two worlds: the (usual) visible (*getig*) realm, and the invisible/intangible (*menōg*) realm. The only way that one can see into the *menōg* realm before death is through drinking *haoma*. It rendered the consumer of this "liquid, omniscient wisdom" *stard*, meaning as if stunned or dazed by a blow, or as being "sprawled on the ground"; in which condition, resembling sleep, visions of what is believed to be a spirit existence could be seen. Flattery (ibid.) also discusses the blissful and visionary nature of the *haoma* experience, and the request of the drinker for "straightness" of mind when confronted with an "ordeal" of encountering truth or justice.

THEORIES FOR THE BOTANICAL IDENTITY OF SOMA/HAOMA

It seems apparent that many of the theories that have been proposed for *soma/haoma* can be readily eliminated. Firstly, the passages cited from the *Ṛgveda* to what seems to be visionary or entheogenic experience cannot be the result of meditation, sleep-deprivation or fasting alone, as the poets clearly state that they have drunk *soma*. Secondly, if it is accepted that the consumption of *soma/haoma* resulted in a 'visionary' experience, then the non-psychoactive plants, particularly the *Sarcostemma brevistigma* or *Periploca aphylla* vines that are sometimes used in contemporary Vedic ritual (and generally known to be substitutes for *soma*) can be ruled out. Various kinds of alcohol have been proposed, but this seems to be improbable owing to the distinctions made in both Zoroastrian and Vedic texts between 'base' alcohol and 'elevated' *soma/haoma*. Cannabis has been proposed by several theorists but also seems highly unlikely due to its relatively mild effects and also the processes described in the *Vedas* to extract the juice of the *soma* stalks.

Gordon Wasson's publication in 1968 of *Soma: Divine Mushroom of Immortality* for the first time introduced the idea of *soma* being a psychedelic, in the form of the red, speckled fly agaric mushroom, *Amanita muscaria*; this theory still enjoys support from several scholars. However, Wasson himself failed to have an ecstatic experience on fly-agaric; on numerous occasions during 1965 and 1966 he ingested the mushrooms but usually felt nauseous and fell asleep. The most active psychoactive compounds in *Amanita muscaria* are muscarine and muscimol, which are quite toxic, and apart from a few exceptions, most experimenters do not have an experience that is particularly insightful, visionary or ecstatic. McKenna (1992) reports twice consuming fly agaric; he felt nauseous and experienced stomach cramps and blurred vision. Beyer (2010) remarks that although small doses of *Amanita* can produce euphoria...

> *"doses large enough to cause hallucinations – which appear to occur only rarely and sporadically – are physically incapacitating, with*

effects including drowsiness, confusion, muscle twitches, loss of muscular coordination, and stupor."

Brough (1971) published a detailed review of Wasson's thesis. One of the strongest objections is that in Siberia the mushrooms are either consumed whole or, more commonly, only the caps are eaten. Brough also dealt comprehensively with Wasson's erroneous interpretation of Vedic passages that led Wasson to claim that 'purified' urine containing fly-agaric alkaloids was drunk in Vedic ritual. Wasson's textual evidence is almost entirely from passages in the *Ṛgveda* and he ignores the detailed descriptions, particularly in the *Yajurveda* and *Śatapatha Brāhmaṇa*, of the elaborate pounding with stones of the *soma* plant. The dried mushrooms could be sprinkled with water to inflate them, but why would they need to be extensively pounded? Also, Falk (2003) cites several references in brahmanical texts that expressly prohibit the consumption of mushrooms.

Currently, the most widely supported scholarly theory – endorsed by Falk (1997; 2003), Kashikar (1990), and Nyberg (1997) – is that *soma/haoma* is the ephedra plant, of which there are around 40 kinds; it is known as *mā huáng* (yellow hemp) in China, where it is used medicinally. The ephedra bush grows widely in the Mediterranean region, the Middle-East and China. Its branches can be easily cut into stalks. It is used medicinally to improve muscle strength and alleviate fatigue, and for asthma and low blood pressure. It is a stimulant that can be toxic in high doses of a concentrated preparation (Spinella, 2001). Falk's (1989; 2003) case for ephedra includes the local identifications in Afghanistan, Baluchistan and northern Pakistan of the ephedra plant as *hōm* (or *hum/huma* or the like); that it is used as both a medicine and as a stimulant (which would fit Vedic descriptions of the *soma* plant being *jāgṛvi*, 'keeping alert'); its extracted juice is bitter (or 'sharp') and reddish; it has a stem, roots and branches; it can be aphrodisiac; and it can cause vomiting; it is used by Zoroastrians in Iran and India, and until recently (1975) by Numbudiri brahmans of Kerala in their *soma* rituals.

Although the stimulating effects and many of the other properties of *soma/haoma* are widely agreed upon, what remains as contentious is whether or not *soma/haoma* had psychedelic or entheogenic properties. Both Falk and Nyberg see no evidence that *soma/haoma* was necessarily psychedelic or 'visionary'; it was only stimulatory. However, Flattery and Schwartz (1989) argue that the pharmacological intensity of ephedra is too weak for it to have been the plant used in ancient Iran. They point out that Zoroastrian priests have been drinking ephedra extracts for centuries without noticing its intoxicating effects; that ephedra is unknown in Indic or Iranian folk medicine; and that it is not regarded as intoxicating in China. I would add that there is nothing like 'reverence' for ephedra in any culture that currently uses it; and neither Falk nor Nyberg take into consideration what *mada* would seem to indicate in Zoroastrian and Vedic texts.

Flattery and Schwartz (1989) present an elaborate case for *haoma* being Syrian (or 'mountain') rue (*Peganum harmala*), reviving a proposition first made in 1794 by William Jones. It has been used for centuries as a medicine and an intoxicant in Iran and neighbouring regions; its seeds are still used for apotropaic purposes (ibid.); and it is used as an aphrodisiac by Turks and others (ibid.). However, Furst (1976) remarks that although known to Arab physicians since antiquity for its intoxicating potential, so far as is known, *Peganum harmala* has never been employed as a hallucinogen. Syrian rue contains almost equal quantities of harmine and harmaline. Harmaline induces a state of relaxation and a tendency to withdraw from the environment, to keep eyes closed, and to want all noises and sounds to be kept to a minimum (Naranjo, 1973). It lowers blood pressure and may also induce sleep. There is, however, some evidence that large doses of harmaline, even though liable to cause nausea and vomiting, can produce visionary experiences that could almost be classified as psychedelic (Beyer, 2010).

For intoxication with *Peganum harmala*, it is usually the angular, red/brown seeds that are used as they have the highest content of harmaline and harmine, which are β-carboline derivatives and monoamine oxidase (MAO) inhibitors. The seeds are usually boiled before being crushed, producing a reddish extrusion. A problem with the mountain rue thesis is that neither in the *Avesta* or the *Veda*s is *soma/haoma* mentioned as being seeds: rather, it is consistently referred to in both sets of texts as what seems to be fibrous twigs or stalks. Nevertheless, it is well known to anyone who has investigated ayahuasca that if *Peganum harmala* is mixed with any of the dozens of plants now known to contain N,N-dimethyltryptamine (DMT), a powerful psychedelic concoction can be produced.

SOMA AS 'ANAHUASCA'

In South Asia, by the time of the early *Brāhmaṇa*s (from c. 800 BCE) the 'original' *soma* appears to have become rare, if not unavailable; though this is not certain. The *Śatapatha Brāhmaṇa* (4.5.10.2–6) mentions several substitute plants—should *soma* be unavailable or stolen—that may be pressed for *soma*, including brown *dūb* (= dūrvā) grass (which is said to be akin to *soma* plants), and yellow (or greenish) *kuśa* grass. Dūrvā is identified by Gonda (1985) as *Cynodon dactylon* (= *Panicum dactylon* = *Agrostis linearis*), which is known, amongst other names, as couch grass or Bermuda grass. *Kuśa* is usually but not always identified by both ancient and modern authors as the same grass as *darbha* (Gonda, 1985), often as the common grass *Eragrostis cynosuroides* (= *Poa cynosuroides*), which grows ubiquitously in north India. Other substitutes are mentioned in *Brāhmaṇa*s and *Śrautasūtra*s, including the juice of *pūtīka*s (Kane, 1997), which are also probably grasses (Houben, 1991). Deeg (1993) maintains that in older Vedic literature *darbha* grass is regarded as an intoxicant, though he does not explain how this common grass could have produced such an effect. The solution being proposed is that *kuśa* (or *darbha*) grass, being related to phalaris grass, most probably contains DMT.

Other substitutes for *soma* mentioned in the *Brāhmaṇa*s include *muñja* grass, *kattṛṇa* (a kind of grass), and *parṇa* (= *palāśa*), which is *Butea frondosa*, a large-leafed tree with orange flowers. The *Aitareya Brāhmaṇa* (5.7.28–32) recommends for *kṣatriya*s the pressed tendrils of the *nyagrodha* (banyan) tree and the fruits of the *udumbara* (*Ficus glomerata/racemosa*), aśvattha (*Ficus religiosa* = peepal) and *plakṣa* (*Ficus infectoria*) trees.

The proposition being made is that *soma/haoma* was entheogenic, which seems to be indicated by the texts (though some scholars dispute this); and that ancient Vedic and Zoroastrian ritual might plausibly be conceived as originally developed primarily as ritual vehicles for an entheogenic experience. It has been argued that mushrooms do not fit the textual descriptions of pressing processes, and that ephedra is insufficient on its own to elicit entheogenic experience.

The suggestion being made is that *soma/haoma* was in all probability never a single plant but a combination of plants. In South America ayahuasca is usually made from the *Banisteriopsis caapi* vine, which provides the MAO inhibitor, and the leaves of *Psychotria viridis*, which provides the DMT. In Asia, various kinds of grasses, particularly *darbha* or *kuśa* grass could have supplied the DMT, which remains orally inactive unless mixed with an MAO inhibitor. My supposition is that *Butea frondosa, Ficus glomerata, Ficus religiosa* and *Ficus infectoria* may contain an MAO inhibitor (I am currently in the process of trying to get these plants analysed). If this is indeed the case, then it would seem that in the late Bronze Age there was sufficient botanical knowledge to manufacture a form of ayahuasca (or 'anahuasca'). The chemistry of ayahuasca analogues was, I believe, the basis of the concoction, to which other psychoactive plants, such as cannabis and ephedra, were probably added by some groups of those who prepared the *kykeon*, the 'mixed potion' of *soma/haoma*.

Avesta: Hōm Yasht: Josephson, J., (1997). *The Pahlavi Translation Technique as Illustrated by* Hōm Yašt. Uppsala Universitetsbibliotek, Uppsala, Sweden.

Beyer, S. V. (2010) [2009]. *Singing to the Plants: A Guide to Mestizo Shamanism in the Upper Amazon*. University of New Mexico Press, Albuquerque.

Boyce, M., (1975). *A History of Zoroastrianism*, Vol. 1. E. J. Leiden/Köln, Brill.

*Brāhmaṇa*s: *The Aitareya and Kauṣītaki Brāhmaṇas of the Rigveda*, trans. Keith, A. B., (1925). Harvard University Press, Cambridge.

Eggeling, J., ed. and trans., (1995) *Śatapatha-Brāhmaṇa, according to the text of the Mādhyandina school* (parts I–V), [1882–1900]. Motilal Banarsidass, Delhi.

Brough, J. (1971). Soma and *Amanita Muscaria'. Bulletin of the School of Oriental and African Studies*. Vol. XXXIV, part II, 331–362.

Deeg, M., (1993). Shamanism in the Veda: The *Keśin*-Hymn (10.136), the Journey to Heaven of *Vasiṣṭa* (ṚV. 7.88) and the *Mahāvrata*-Ritual. *Nagoya Studies in Indian Culture and Buddhism (Saṃbhāṣā)*, Vol. 14, 95–144. Department of Indian Philosophy, University of Nagoya.

Elizarenkova, T., (1996). The Problem of Soma in the light of language and style of the Ṛgveda. In Balbir, N., & Pinault, G. (eds.) *Langue, style et structure dans le monde indien (Centenaire de Louis Renou)*. 13–31. Paris: Editions Champion.

Falk, H., (1989). Soma I and II. *Bulletin of the School of Oriental and African Studies*, Vol. LII, Part 1, 77–90.

Falk, H., (2003). Decent Drugs for Decent People: Further Thoughts on the Nature of Soma. *Orientalia Suecana*, Vol. LI–LII, 2002–2003, 141–155.

Flattery, D. S., Schwartz, M., (1989). *Haoma and Harmaline: The Botanical Identity of the Indo-Iranian Hallucinogen "Soma" and its Legacy in Religion, Language and Middle Easter Folklore.* University of California Press, Berkeley/Los Angeles/London.

Furst, P. T. (1976). *Hallucinogens and Culture.* Chandler & Sharp Publishers, Inc., San Francisco.

Gershevitch, I. (1974). 'An Iranianist's View of the Soma Controversy'. Gignoux, P. & Tafazzoli, A. (eds.) *Mémorial de Jean Menasce*, .45–75. Imprimerie Orientaliste, Louvain.

Gnoli, G. (2005) [1987]. 'Haoma'. In *Encyclopedia of Religion*, Jones, L., (ed.). Vol. 6, 2nd edn., 3775–3776. : Thomson/Gale, USA/London/Munich.

Gonda, J. (1975). Vedic Literature (Saṃhitās and Brāhmaṇas). Gonda, J., (ed.) *A History of Vedic Literature*, Vol. 1. Otto Harrassowitz, Wiesbaden.

Gonda, J. (1982). *The Haviryajñāḥ Somāḥ: The Interrelations of the Vedic Solemn Sacrifices; Śāṅkhāyana Śrautasūtra 14, 1–3, Translation and Notes.* North Holland Publishing Company, Amsterdam/Oxford/New York.

Gonda, J. (1985). *The Ritual Functions and Significance of Grasses in the Religion of the Veda.* North Holland Publishing Company, Amsterdam/Oxford/New York.

Hillebrandt, A. (trans. Sarma, S. R.) (1980–1981) [1927–1929]. *Vedic Mythology*, Vols 1–2. Motilal Banarsidass, Delhi.

Houben, J. E. M. (1991). *The Pravargya Brāhmaṇa of the Taittirīya Āraṇyaka: An Ancient Commentary on the Pravargya Ritual.* Motilal Banarsidass, Delhi.

Houben, J. E. M. (2003). 'The Soma-Haoma problem: Introductory overview and observations on the discussions'. *Electronic Journal of Vedic Studies [EJVS]*, Vol. 9, Issue 1a (May 4).

Kane, P. V. (1977–1997). [1930–1962]. *History of Dharmaśāstra*, Vols. 1–5, 3rd edn. Bhandarkar Oriental Research Institute, Poona.

Kashikar, C. G. (1990). *Identification of Soma.* Tilak Maharashtra Vidyapeeth, Pune.

McKenna, T. (1992). [1984]. *Food of the Gods.* Rider, London.

Modi, J. J. (1922). *The Religious Ceremonies and Customs of the Parsis.* Mazagaon. British India Press, Mumbai.

Naranjo, C. (1973). *The Healing Journey: New Approaches to Consciousness.* Ballantine Books, New York.

Nyberg, H,.(1997). The problem of the Aryans and the Soma: The botanical evidence. Erdosy, G. (ed.), *The Indo-Aryans of Ancient South Asia: Language, Material Culture and Ethnicity*, 382–406. Munshiram Manoharlal, New Delhi.

O'Flaherty, W. D. (1969) [1968]. The Post-Vedic History of the Soma Plant. Wasson, G. (1969). *Soma: Divine Mushroom of Immortality.* 95–147.

Parpola, A. (1997). The problem of the Aryans and the Soma: Textual-linguistic and archeological evidence. Erdosy, G. (ed.), *The Indo-Aryans of Ancient South Asia: Language, Material Culture and Ethnicity*, 353–381. Munshiram Manoharlal, New Delhi.

Spinella, M. (2001). *The Psychopharmacology of Herbal Medicine: Plant Drugs That Alter Mind, Brain and Behavior.* The MIT Press, Cambridge, London.

Staal, F. (2001). 'How a Psychoactive Substance Became a Ritual: The Case of Soma'. *Social Research*, Vol. 68.3, 745–778.

Sturhman, R. (1985). 'Worum handelt es sich beim Soma'. *Indo-Iranian Journal* 28, 85–93.

Sturhman, R. (2006). 'Capturing Light in the Ṛgveda: Soma seen botanically, pharmacologically, and in the eyes of the Kavis'. *Electronic Journal of Vedic Studies [EJVS]*, Vol. 13, Issue 1 (April), 1–93.

Thompson, G. (2003). 'Soma and Ecstasy in the Rgveda'. *Electronic Journal of Vedic Studies [EJVS]*, Vol. 9, Issue 1e (May 6).

Wasson, G. R. (1969) [1968]. *Soma: Divine Mushroom of Immortality.* Harcourt Brace Jovanovich, Inc., New York.

Vedas:

Hymns of the Atharvaveda, Vols 1–2; *The Hymns of the Ṛgveda*; *Hymns of the Sāmaveda*; *Texts of the White Yajurveda* (1985; 1999; 1986; 2012) [1895–1896], trans. Griffiths, R. T. H.). Munshiram Manoharlal, New Delhi.

Rig Veda Samhitā, Vols 1–12 (ed. and trans. Kashyap, R. L.) (2007–2009). Sakshi, Bangalore.

The Yajur Veda (Taittirya Sanhita) (ed. and trans. Keith, A. B.) (2008) [1914]. Forgotten Books, New Delhi.

AYAHUASCA, ECOLOGY AND INDIGENEITY

WILLIAM ROWLANDSON

What do we talk about when we talk about ayahuasca? What questions are thrown up? What problems, dilemmas, reflections, solutions? What does ayahuasca *mean*?

Ayahuasca has grown tremendously in just a few years. My first encounter with it was as *yage* in the exchange of letters between William Burroughs and Allen Ginsberg. Burroughs set out to Central and South America in 1953 on the scent of a little-known brew that enabled telepathy: 'When I started looking for Yage I was thinking along the line the medicine men have secrets the whites don't know about' (2006). Ginsberg, drinking ayahuasca in Peru seven years later, presents its allure as a lost alchemy, a poetic secret. There is an intrepid, trailblazing tone to their letters.

In the summer of 2013 filmmaker Michael Wiese, who first encountered ayahuasca as a potential cure for his Parkinson's (a very successful one, it turns out), said to his audience: 'all of us here – having drunk ayahuasca or not – have been touched by ayahuasca'. Compelling words, I felt at the time, stretching beyond an expectantly sympathetic (and enthusiastic) audience.

There is clearly something meaningful about ayahuasca, some resonance, some worth. Ayahuasca winds through our world at many levels, energising ideas, ethics, politics and visions. I find it significant that a bitter brew from the Amazon touches so many areas that seem so relevant in this jagged modern age. Ayahuasca is not a technology we are familiar with. It is a science unlike what we are taught about science. It is a medicine not prescribed by doctors, not concocted in a lab. It is an unlikely source of knowledge and experience.

When people talk about ayahuasca, they tend to ask questions that I am in the habit of asking. They consider things that they consider worthy for consideration; which are things that I also consider worthy for consideration. I consider them important. I am therefore happy that this interest in ayahuasca has grown as I continue to be fascinated by the implications of its growth and the conversations that arise.

There is a sense of *enchantment* about ayahuasca that we are encouraged to grow out of as we leave childhood. Wizards, Indians, magic potions and witches' brews, spells and incantations, strange beasties in the jungle. But these are fairy stories! Adventures in Wonderland. Stories of good witches and bad witches who send spells to each other. The hero who is helped by a talking cat. It is an enchantment that animates and stirs the imagination. It creeps through our solid structures of reality, compelling us to consider the structures from a slightly altered perspective.

Ayahuasca is referred to as a female, feminine, form. I likewise perceive her as feminine. Feminine wisdom. She is often referred to as a blue spirit or a serpent, even to the suspicious Burroughs: 'and then this blue spirit got to me and I was scared and took some codeine and Nembutal' (2006). She is experienced as a counterpoint to some of the more aggressive and destructive aspects of our culture – yet she can also be aggressive and destructive. She challenges some of our most basic assumptions about nature – about reality. The experience of a wise and conscious universe. A different vision of the world. A strange and alluring magic.

And yes, ayahuasca is a drug. Of course it is a drug. It is a drink made from plants and water. No faith required. No credo. Just drink it, as Terence would say, and then make up your mind. That is the power. There is a long tradition in the technology of brewing. It is a subtle chemistry. There is also a long tradition of navigating the physical and mental landscape of the drink, anticipating its effects on the body and the imagination.

Yet as a drug it commands a particular language. Ayahuasca is not sold in pill form, nor bought in a can at the corner store, nor smuggled into nightclubs. It is rare to read an account of someone chowing on ayahuasca at home on a Friday night after the pub. Not a club drug. She requires investment.

I have not drunk ayahuasca, and for all I know the assumptions that I have made about ayahuasca will change when I do drink ayahuasca. I already assume that I should not assume too much about what I already assume about ayahuasca. And even *that* assumption might change…

So ultimately I am not saying what ayahuasca means. I would not presume to know what she means to other people. I am saying what ayahuasca means to me, how what ayahuasca means to me seems to chime with what she means to others, whether they have drunk or not.

Boundaries are crossed. In order to drink ayahuasca international borders may need to be crossed. Travel, insurance, accommodation must be booked. Time is required. Preparation – perhaps the *dieta*. The trip begins early and requires attention. And for the most part the destination is the Amazon, an environment generally different from home. The forest is deep and the river has more traffic than the roads. Colours are intense and much is unfamiliar. This is abundantly clear from the *Yage Letters*; the experiences of drinking the brew are slight in comparison with Burroughs' picaresque adventures in Panama, Colombia and Ecuador, his scrapes and tussles, sexual encounters and early explorations into the *Interzone* of his own roguish mental landscape.

The location of the drinking may be a building open to the air yet enclosed, never distant from trees and smells and noises. It is exotic. Things are manifestly different – not fewer things but different things. A group gathers. Trust of strangers is thus required, and trust in particular in the *ayahuasquero*. Yet there is an invitation to trust, as the tradition of ayahuasca is old and legitimate. It is not illegal. The cops won't bust down the door. The circle is made to be as safe as possible. The setting of the gathering will have a large influence upon the experience. It cannot be removed from the experience. The setting is the experience.

Things are revealed – the body and the mind are not separate as the nausea and vomiting may come from deep within the psyche. The chants or rattles become physical, visual, material. An environment that is already strange becomes radically stranger. Further boundaries are crossed. Some kind of dialogue unlike a day-to-day dialogue occurs, an internal dialogue, within the group, with the *ayahuasquero*, with the trees and animals, with the visions behind closed eyes.

Every experience is recounted in a different way but there is a commonality to the descriptions that I feel immediately compelling. It is a sense that things are not quite what they normally seem. There is an invitation to experience a state of existence quite unlike other states, to consider the value of this state of existence beyond the night of the drinking, to recognise that the forces experienced are operational throughout our lives, not just in the mode of altered perception. Harrowing or joyful, there is a consensual sense that the blooming, buzzing confusion experienced on a night in the jungle is not a different reality, but a part of our reality that we are trained to dismiss as a different reality, as a fantasy reality. That which is fantastic is no *mere* fantasy. Life is a hero's journey, a fairy tale.

Thus ayahuasca and other plants and plant preparations are known as teachers. Something transformational occurs with the greater experience of ayahuasca that activates a re-evaluation of some foundational principles upon which our industrialised modern societies are based. This is the central focus of Ralph Metzner's stirring introduction to *Sacred Vine of Spirits: Ayahuasca* (2005), and it is a sentiment reflected in accounts of ayahuasca experiences. Again, Burroughs puts it in his own sardonic style:

> *"Yage is it. It is the drug really does what the others are supposed to do. This is the most complete negation possible of respectability."* (2006)

Green pervades. No account that I have read or heard fails to evoke a profound vegetal dimension to the experience – a sense of astonishing interpenetration of human and plant biology. Dennis McKenna (2012), for example, describes an experience after drinking ayahuasca of entering the body of the trees, getting smaller and smaller – cellular, molecular – until passing through the very process of photosynthesis. Such a profound experience cannot be separated from a sense of our relationship with the natural world in which we live. It is a botanical experience. It is an ecological experience.

This seems to me an integral aspect of the whole paradigm of ayahuasca; at least, this is the narrative that sings to me from the numerous accounts. I am fascinated by the growth of ecological and environmental discussions and activities that are sprouting from the experience of ayahuasca. There is, I feel, a tremendous value in this. As Metzner indicates, the experience of ayahuasca and other plant teachers problematises many of the assumptions upon which our cultures are based. Ayahuasca interrogates the long-held assumption that the natural world is not conscious, that any meaningful dialogue with the landscape is at best romantic, at worst pathological; that natural resources are for the taking; that to be human is to participate by default in massive environmental damage. Ayahuasca thus questions the concomitant ease with which rich, thriving ecosystems of great biodiversity may be reduced to sterile wastelands in the pressure to obtain resources. Terence McKenna often suggested that psychedelics (especially tryptamines) were tools to help us find a way out of the mess. This is a possibility with ayahuasca, as it emerges from the very biological, geographical and ethnic contexts that are under greatest threat from modernity.

I am drawn to these narratives because I genuinely consider plant-induced altered states of consciousness in the woods at night to be amongst the most important and transformative events of my life. I have invested myself deeply into the state of being that is the woodland. I have evoked the spirit of ayahuasca whilst gazing at the dark trees flickering in the fire light. What do I know of her? What has she said to me through those who have drunk her and described their experiences? What vision of the world is described that parallels the visions of the world that I describe from my own experiences?

The true impact of this involvement with trees came whilst I was writing my latest book, *Imaginal Landscapes* (2014). This woodland nature is how Swedenborg presents human existence – indeed all existence. Life and death grow in and out of each other in such a way that life is death and death is life. The image of the tiny acorn sprouting into a mighty oak is challenged – nothing to everything then back to nothing – as the woodland presents a different picture. A tree falls and the branches become vertical – they are now trees. A tangle of roots exposed when it fell slowly develops into up-growing trees. Ancient ivy around an oak becomes oaken, incorporated into the oak's massive structure. These rigid divisions are challenged by Swedenborg and by night-time magic in a woodland. It is a vision that challenges our most deeply-held understanding of the dynamic opposition of life and death. Oblivion to existence to oblivion. The *imaginal*, as so poetically related by James Hillman, is where fact and fiction, reality and fantasy, life and death, dream and waking, all swirl together in a baffling state of enchantment.

I find a relationship here with Patrick Harpur's book *Daimonic Reality*. He makes his own distinction between Spirit and Soul. Spirit is the urge towards perfection – the impulse towards light and clarity. It is a sensation traditionally evoked by monotheism. Christhood as perfection. It is a masculine, Saturnine principle, guiding the individual and the collective onwards, upwards in a trajectory of yearning. Such a driving principle can translate into the spirit of progress, of repeated waves of modernity. This is our cultural paradigm. Woodlands are cleared of fallen branches; the forest floor is sterile. Mountains are valued not as the dwelling-places of ancestors, but as repositories of minerals. The forests are immensely valued, but as a resource, or as a layer of clothing to be stripped to reveal the rich oil beneath the soil. We place immense value in the natural world, but a value such as Smaug places in his treasure hoard, and like Smaug we fight tooth and nail to own such commodities. Our Western culture is one of Light. It is the principle upon which we judge wisdom, comfort and safety. To be enlightened is to see the light, to see the truth, to banish the shadows. No dark corners remain in our houses. The streets at night are ablaze. The stars are hidden in the loom. We are safe from the unknown. This is the spirit of Spirit.

Yet such a spirit accommodates only uneasily the Soul. Soul is feminine and androgynous, mischievous, mercurial, unclear, dark, compellingly un-perfect. Soul is undervalued, rejected, cast aside. Soul ripples in the back-eddies of moonlit waters. Soul is pagan, daimonic, sensual and sexual.

The story of ayahuasca is, I feel, a soul-story. Ayahuasca is witchy bitchy river water swirling mysteriously, muddy and smoky in a battered plastic bottle. Visible songs guide the drinkers. The old Indian with a Pepsi T-shirt, denim shorts and flip-flops blows tobacco smoke over the bubbling brew, adding beakers of murky water, adding sticks to the fire. Hidden memories spring from the darkness; shadows emerge from forgotten corners of the psyche. The trees vibrate with magic. Boundaries are blurred. Is the tree living or dying? The roots of all these trees are enmeshed below ground like the branches above. Everything is alive! Are the leaves on that tree waving at me? Are those patterns in my vision illusion or something more than illusion? Is the whole experience in the jungle an illusion?

Attention is focused on the trees. Trees are felled. Huge swathes of forest are cut down. This is horrible, wrong. How dare they? How dare we? How dare I? Am I contributing to this destruction? Yes. How can I not contribute? What system do I belong to, and what power do I have within that system to reduce the destruction? Am I destructive? Yes. To whom? To family, friends, strangers, myself. I fell trees. We all fell trees. I destroy. We all destroy. Life is destruction. How can we limit the damage?

Ayahuasca is not only called a teacher, she is a stern teacher. Most accounts describe such dialogues of self-evaluation. Psychology becomes ecology. Ecology becomes ethics. Any system of ethics that does not hold ecology, environment, the earth, at its core becomes sham. A crazy mirror is held up to the drinker, reflecting grinning monsters and demons, inspiring a dialogue between the self and the self, a scrutiny of the self's relationship with the community, with society, with the activities and principles that are central to our society. The soul-journey of ayahuasca is thus an investigation into the values of Western culture. As such, the whole experience of ayahuasca – this shamanic brown jungle brew – is an experience of the idea of *indigenous*.

I have often asked my students to discuss their understanding of indigenous. What do they think about when they hear or use the term? Who is indigenous here? Who is not? Why not? There is expected divergence of opinion; yet there is also consensus. It is an important question, as a consideration of the indigenous is a consideration of fundamental questions of identity – our relationship with the land, with the past, with each other.

'Indigenous' is a bundle of values corresponding to absence. To list the values associated with indigenous is to identify qualities lacking in the non-indigenous. This is the distinction far more than ethnicity, race, language or geography. As such, and especially present within the extensive literature of ayahuasca, indigenous represents the desire to incorporate the values embedded within the term indigenous. How do *they* live? What do *they* do that which we do not? How can we learn from them? Indigenous represents what is absent from Western culture to the extent that the term 'Western' means, precisely, that is *not* indigenous.

This is a delicate matter. My colleague David Stirrup illustrates the tendency amongst white European political groups to appropriate the term indigenous in order to justify ethnic supremacy and anti-immigration rhetoric. Indeed the figure of a Native American warrior in full headdress has been used in Sweden and Switzerland to vindicate this far-right agenda of 'native' people resisting the settlers. Stirrup explains why the BNP are not 'Indigenous rights activists':

> *"That aspiring political representatives in one of the most successful colonizing nations in the history of the world should demand protection from the fruits of its own success is, to say the least, highly ironic."*
> (2013)

He argues with regards to the Finnish disrespect for the indigenous Sami people that 'the Far Right's interest in indigeneity begins and ends with self-preservation' (2013). The protections afforded by the legal definitions of indigeneity risk being undermined by the appropriation of the term by non-indigenous communities such as the BNP. Whilst far-right groups may demand recognition of their own indigeneity, such an appeal impoverishes the very meaningfulness of the term for, precisely, indigenous people.

So I am not asking why I am not indigenous. I am interrogating the dazzling array of values associated with the term. I am deliberately not avoiding generalisations, as the terms 'Western' and 'indigenous' are unavoidable generalisers. What are these values?

Indigenous signifies people who *belong*. Theirs is a community, whilst ours is fragmented society. Theirs is a connection to the land, an innate sense of ecology, a vision of balance and health. Indigenous is authentic. It resists the dominating culture of Gringoland. They are colonised not colonisers, oppressed not oppressors, victims not victimisers, natural not artificial, instinctively spiritual not religious. Their homes, lands and lives are threatened by loggers, cattle farmers, mining companies, *petroleros* and corporations, who are our ambassadors in their lands. They value tradition, and their ancestors are present in dreams and visions. They still respect the crazy shaman. Their doctors are healers. Their medicine is ritual. They value the living forest, rivers, plains and mountains, not the gimmicks, gadgets and plastic toys of Western culture. They have something we lack. Indigenous is not Western because Western is not indigenous. The terms are value-filled and have grown to evoke opposing value systems.

Of course this is not accurate. The Inca were a dominator society, creating a vast empire on the scale of the Roman. Moctezuma's Aztecs were a bloodthirsty lot. Indigenous people cut down trees and are nasty to each other, have mobile phones, drink Coke and eat burgers. There are *ayahuasqueros* who drink the brew and do great mischief, or who provide deliberately dodgy brews. There are scoundrels and bastards in all cultures and all societies. Five centuries of colonial and corporate abuse of indigenous cultures, however, reveal that Western culture has built scoundrelry and bastardness into its foundational principles.

Furthermore, the values that are ascribed to the term indigenous may not be present in indigenous cultures nor recognised as values. What we may perceive wistfully as the moral attribute of sharing may be a simple strategy for survival. The ethic of harmonious living with the environment again may not be recognised as an ethic at all, but as a straightforward necessity to guarantee natural resources for the community.

Drinking ayahuasca does not make one indigenous, no matter how many times one returns to the lodge in Iquitos. What prevents me being indigenous is, crucially, the fact that I am not indigenous. Neither am I Russian nor a woman. And yet we may consider the bundle of values associated with *our* understanding of indigenous, and question whether we cannot attempt to embody those values. This contemplation of values may, therefore, urge us away from a pretence at indigeneity either for belligerent political agendas – like the BNP – or for fashion agendas, such as wearing 'ethnic' clothing so as to seem more groovy. Are those values that we identify in others (appropriately or not) achievable? This, to me, is the heart of the matter.

Ayahuasca is deep immersion in indigeneity, which is thus the radical experience of the Other. Yet rather than mourning our lack of such values, the spirited debates surrounding ayahuasca reveal to me a desire to make those values our own. Thus ayahuasca encourages – at least *should* encourage – a scrutiny of the very activity of drinking ayahuasca. As such I was heartened to participate in an earnest debate at Breaking Convention in 2013 about, precisely, the impact of ayahuasca tourism. Tread gently. Be respectful. What litter are you dropping? What are the consequences of more gringos exchanging dollars for visions? How is the act of drinking part of a problem as much as a solution? Such questions are immensely important, as they bring the act into sharp focus against a backdrop of centuries of injustice and brutality. Is the growth of ayahuasca in the Western mind another act of colonial cultural appropriation? Are the *ayahuasqueros* being lured away from their traditional roles as *curanderos* and healers and active members of the community in order to cook up the brew for strangers? Iquitos was at the heart of the rubber boom a century ago, a period characterised by exploitation of natural resources, contamination of the ecosystem and abuses of the population. Iquitos is now at the heart of the aya boom. How different is the process? How can drinkers respond to such intense questions and yet still receive the gift of aya visions? Can the act of drinking ayahuasca mitigate the act of drinking ayahuasca by compelling the drinker to question the act of drinking ayahuasca?

I am emboldened by discussions of the spirits of the woodlands, as it is a language I have secretly spoken all my life. I have not experienced in a dazzling vision the female form of ayahuasca, and yet I have evoked her spirit and have meditated long with her. Slowly, vine-twistingly, an experiential magic that has long been mistrusted, vilified, pathologised and criminalised in our thrusting brave societies is returning. It *is* possible to belong, to be part of the land, the community, to respect the muddy rivers, snake-eyed trees and ancient societies of our own homes, to respect magic and the ancestors, to heal. Above all to heal.

Burroughs, W. S., & Ginsberg, A. (2006). *The Yage Letters Redux*. City Lights Books, San Francisco.

Harpur, P. (1995). *Daimonic reality: a field guide to the otherworld*. Arkana, London.

McKenna, D. J. (2012). *The Brotherhood of the Screaming Abyss*. Polaris Publications, Minnesota.

Metzner, R. (2005). *Sacred Vine of Spirits: Ayahuasca*. Inner Traditions, Vermont.

Rowlandson, W. (2014). *Imaginal Landscapes*. Swedenborg Society, London.

Stirrup, D., & Padraig, K. (2013). 'I'm indigenous, I'm indigenous, I'm indigenous': Indigenous Rights, British Nationalism, and the European Far Right. Mackay, J. & Stirrup, D. (eds.) *Tribal Fantasies: Native Americans in the European Imaginary*, 1900-2010, 59-84. Palgrave Macmillan, Basingstoke.

THE POLITICS OF ECSTASY: THE CASE OF THE BACCHANALIA AFFAIR IN ANCIENT ROME

CHIARA BALDINI

"To consider nothing wrong was the highest form of their religious devotion."
Livy

Second Century BC. It is late at night in Rome. Small groups of people are silently walking in the streets of the capital of the Empire. They bear torches in their hands, lighting their way and their faces; blossoming young women and experienced matrons, young free men and slaves, members of important families and simple people, citizens of Rome and people coming from all over the Empire... they disappear behind the doors of private houses.

When they have all arrived, the music starts. The flutes weave an intricate melody filling the rooms with their high-pitched tones;

percussionists soon join in with their deep vibrations, designing compelling rhythm patterns. The sound builds up... faster and faster... louder and louder... the participants start dancing, letting themselves be moved by the music, while the chants spontaneously flow out of their bodies, as the powerful rhythms keep in pace with the music. Someone starts screaming in long loud shrieks, followed by many others. They soon find themselves dancing frenetically to the music... the cries are so wild that they don't seem human anymore. Progressively, the sound of enthusiasm gives way to that of pleasure... powerful release... deep liberation... ecstasy... nature taking possession of the bodies, breaking the barriers of conventions, the powerful healing power of pleasure weaves its magic... while the night all around unfolds its mystery into the morning and the music gradually wanes. At the break of dawn, someone is still singing, others are sleeping... the bodies rediscover their limits, while the initiates into the *Bacchanalia* stumble back onto the streets, shielding their eyes from the light of the sun, holding each other against falling... exhausted, happy, reborn.

In the Second Century BC, the Roman Empire extends its supremacy on both the eastern and western shores of the Mediterranean, arguably becoming the most influential power of the ancient Western world. The capital of the Empire is one of the greatest urban settlements, with almost a million citizens. Etruscans, Campanians, Phoenicians, Egyptians, Greeks, Macedonians, and Persians, all live together with the Romans in one of the very first examples of a cosmopolitan city. Latin and Greek are the 'international' languages and different religious traditions are celebrated freely, as long as they have been sanctioned by the Senate's legislations. *"Pax deorum, pax hominum"* is the motto of the Roman Senate: when the gods are in peace, the men are in peace too. But soon, that carefully controlled peace will be shaken up by the introduction of an enticing foreign cult.

As a result of the large-scale migrations into the city following the Hannibalic Wars (Gruen, 1990), peoples from the rural areas of Latium, from Etruria (Tuscany), and from Magna Graecia (Southern Italy) had swarmed into the capital, bringing with them their traditional religious practices. Before the Senate realised it, a new kind of ritual, called *Bacchanalia*, involving spiritual practices radically different than those traditionally performed in Rome, had spread "with astonishing swiftness" (Gruen, 1990) across the whole city. As opposed to other foreign cults (like the Egyptian cult of Isis or the Middle Eastern cult of Magna Mater), the new rituals had propagated spontaneously, without official approval. Thus Bacchus, the god of ecstasy, wine and wild nature had made his illegal, yet clamorous, arrival at the very heart of the Empire.

Participation in the *Bacchanalia* required the strict observance of a vow of silence concerning what happened during the rituals. As a consequence, we do not have any direct report of the ceremonies, but we do know that they mainly consisted of initiations, which probably involved a ritual death followed by a rebirth into the 'new family' of initiates (Burkert, 1987). We can, however, rely on a fairly large amount of both mysterious and explicit images, which have ignited the imagination of those who came later.

THE JOURNEY TO ROME

But who is Bacchus? And how did his practice arrive in Rome? The *Bacchanalia* were a Roman adaptation of a very ancient spiritual practice, dating back to the Neolithic Goddess culture of Old Europe. These ancient rituals generally comprised of a series of cultic practices involving dancing for many hours to repetitive rhythm patterns, ingesting psychotropic substances (often brews made from different ingredients) and ceremonial sexuality. These techniques, refined during the course of many millennia, were aimed at inducing powerful alterations of perception, states of divine possession, incursions into the world of spirits and ancestors, experiences of mystical union with Nature, the Great Mother (Eliade, 2004; Baring and Cashford 1993).

The prehistoric practices of ecstasy later branched out into Greek culture via the Minoan civilisation of Crete, considered one of the longest surviving examples of Old Europe's "Goddess culture" (Eisler, 1987). The myth and cult of the Cretan Mountain Goddess gradually gave way to the myth and cult of her son, Dionysus, the god of vegetation, fertility, wine, theatre, love and ecstasy; and the Goddess' ecstatic rites thus survived through the cult of her son (Harrison, 1991).

An early form of the newly born Dionysian rituals was reserved for women only (the *maenads*) and was generally celebrated on mountaintops. A more 'urbanised' version, open to men and featuring the characteristics later found in the Roman *Bacchanalia* (dancing, drinking and sexuality), became extremely popular as a 'Mystery Religion' during the Hellenistic era, in the Third Century BC. It was in this phase that the cult became popular in Magna Graecia (Southern Italy) and, from there, it was later exported to Rome. Once in the capital, Dionysus curiously met his Etruscan counterpart, Fufluns, god of wine and wild nature, another remnant of Old Europe's tradition preserved by the very ancient Etruscan culture and imported into the capital by the influential Etruscan populations. Together, Dionysus and Fufluns merged into Bacchus (from Greek *Bakchos*: "frenzied one"), a newly reborn, powerful and extremely alluring god.

BACCHANTES VS. CITIZENS

By the Second Century BC those initiated in the rites of Bacchus reached thousands of people in Rome, constituting a sort of 'parallel city' (Livy in Meyer, 1987) distributed among many powerful private organisations with an independent monetary fund and an internal justice system.

Membership of these associations was open to everyone, including women, foreigners and slaves, which, in the rigidly sectarian, misogynist and clan-based Roman society, made the *Bacchantes* the representatives of a (more or less conscious) counter cultural movement. As a matter of fact, the *Bacchanalia* were rituals where spiritual practice consisted in breaking the very rigid conventions of gender, social status and ethnic origin imposed by mainstream culture. In other words they constituted what we would now call Temporary Autonomous Zones, functioning as privileged spaces for the germination of counter culture, as had been the case already during the Hellenistic phase of the cult in Greece (Cassidy, 1991).

By sharing such revolutionary beliefs and practices the Bacchantes benefitted from an exceptional social cohesion – that which Plato, talking about "Mystery Religions", had defined as "Brotherhood of bodies and souls" (in Burkert, 1987), a spiritual bond even stronger than the political bond linking the initiates to the Roman government. In other words a worshipper of Bacchus was first and foremost a *Bacchante* and, only secondly, a citizen of Rome.

In a heavily militarised society, fully depending on the faithfulness of their citizens to the civic order and values established by the Senate, the vast and illegal proliferation of the *Bacchanalia*, together with its *radicalism*, amounted to a pressing social and political emergency. The Senators did not wait long before finding a way to restore order in the city.

TITUS LIVY

The main source of information concerning the *Bacchanalia Affair*, the name nowadays commonly given to the prosecution in 186 BC of Bacchic worshippers, is the Roman historian Titus Livy. In his *Ab Urbe Condita Libri* ('From the Foundation of the City'), written some 200 years after the facts, Livy gifts us with a very detailed version of the story. It is important to clarify that it was Livy's declared intention to present the history of Rome in terms of a series of moral lessons, to inform, educate and (even) entertain (Coles, 2013). Livy's morality was based on stoic ethics (Riedl, 2012) praising rational control over lust and the repression of desire. As we can imagine, the *Bacchanalia Affair* offered him a perfect opportunity to broadside a perverse activity where the loss of all rational control is celebrated as the peak of religious experience!

In book XXXIX of his work, Livy narrates how consul Lucius Postumius one day happens to witness the confession of the "well known prostitute" Hispala (Meyer 1987). The woman, in an attempt to save her lover and protector Aebutius (a member of the Roman elite) from being initiated into the cult, declares to the consul that she had been once initiated and confesses that the secret rites are "the workshop of corruption of every kind", where "all sorts of enormity would have first to be suffered and then to be practiced", including: promiscuity, group rape of minors, murder of unwilling initiates and forging of the wills of the dead. The consul immediately proceeds to inform the Senate of the extreme threat that lies in the *Bacchanalia*. The Senators, "seized with extreme panic", promptly proceed to promulgate a new set of laws, known as the *Senatus Consultum de Bacchanalibus*, aimed at repressing the cult once and for all.

THE SPEECH

According to Livy, after the Senate composed the *Senatus Consultum*, Postumius called a public assembly to communicate the Senate's decisions to the Romans. This speech (probably taken from the archives of the Postumii family) is generally considered to be the historical base on which Livy wrote his report of the *Affair*, while the first part of the report, concerning the intricate 'soap opera' of Hispala and Aebutius, could have been inspired by a theatre play written shortly after the facts, a common practice to hand down historical facts over the centuries (Walsh, 1996).

Upon addressing the public assembly gathering to be informed on the Senatorial resolutions, Postumius begins:

> *"Citizens of Rome… if I lay bare the whole story I am afraid that I may spread excessive alarm… Whatever I tell you, my words are inadequate to the horror of the situation. Our energies will be devoted to taking the adequate measures."*

Note the fine rhetorical strategy aimed at creating panic even before he introduces the subject. The *Bacchanals*, he explains:

> *"…have for a long time been performed all over Italy… and in Rome itself. I'm sure you are aware of this by the bangings and howlings heard in the night, which echo through the whole city."*

It is interesting to note that today in Latin-based languages 'confusion' and 'loud noise' are still referred to with a word deriving from

Bacchanalia: *baccano.*

"Some believe it's a kind of worship of the gods; others suppose it to be a permitted exercise of playful excess… involving only a few people. But if I tell you that there are many thousands of them you are going to be horrified."

Here we can feel the genuine surprise on the side of Postumius, as a representative of the conservative elite, before a spiritual practice that broke every canon of respectability and moral conduct on which Rome's civic order was based. The high numbers of those who gave up traditional religious conduct to follow the new god had certainly shocked many.

"A great part of them are women, and they are the source of this evil thing; next are males, scarcely distinguishable from females… Dancing frenetically, having lost their minds by lack of sleep, by drink, by the confusion and the shouting that goes on throughout the night."

As a matter of fact, in the *Bacchanalia* (as in the Greek Dionysian and Etruscan rituals) women held leadership positions: they were priestesses and therefore entitled to initiate men into the cult. This was a further invalidation of the patriarchal hierarchy, upon which Roman society was based. Moreover, dance was generally considered unacceptable for Roman men, for it jeopardised their dignity as males and as soldiers.

"Citizens of Rome, do you feel that young men, initiated by this oath of allegiance, should be made soldiers?… Will they take the sword to fight to the end in defence of the chastity of your wives and children?"

The issue of security is used here to pressure public opinion. Postumius makes again reference to male pride: in the moment when other men will come to rape your women and children, will these 'dancing freaks' be able to do anything about it? Moreover he stresses that the *Bacchic* oath could not be compatible with the military oath that young males swore upon joining the Roman army. The two oaths were in fact made exactly at the same age (Limoges, 2009).

"Whatever crime there has been in the past years, you may be sure, has its origins in this one worship… And it is already too serious to be dealt with privately, it requires the supreme power of the State. As a united body we can destroy them."

In the Second Century BC most crimes, including murder, were dealt with privately, according to the 'eye for an eye' principle. Therefore, Postumius' argument that the supreme power of the State was needed to resolve the situation sounded like an exceptional measure to the ears of the Romans. And this is yet another strategy aimed at intensifying the perception of the *Bacchanalia* as an extreme menace.

"Nothing is more deceptive in its appearance than a depraved religion. When the will of the gods is an excuse for criminal acts, there comes to mind the fear that in punishing humans we might be doing violence to something divine… I have thought it right to give you this warning, so that no superstitious fear may agitate your minds when you observe us suppressing the Bacchanalia… *All this we shall do with the favour and approval of the gods."*

Here it has been noticed that Postumius clearly banned Bacchus from the pantheon of the 'accepted gods' (Riedl, 2012), as whatever will be done against his followers will receive the appraisal of the 'good gods' – i.e. those presiding over Rome's civic order.

What follows is the reading of the resolutions of the Senate, known as the *Senatus Consultum de Bacchanalibus*, a copy of which, fully compatible with Livy's version, was found in 1640 in Calabria, southern Italy, as a final confirmation that Livy's report was based on real events:

"The priests of these rites, male and females, are to be sought out not only in Rome but in all market towns and centres of population, so that they should be available for the consuls. No one should ever attempt to celebrate these ceremonies anymore, in Rome or Italy… An inquiry should be held regarding those persons who had assembled or conspired for the furtherance of any immoral or criminal design."

But there was an important exception, marking the Senate's ultimate intention not to eradicate the cult completely, but to impose very strict rules, which would transfer the sphere of action of the worshippers from the private to the public sphere (Takács, 2000):

> "If any person regarded such ceremonies as allowed by tradition and believed him/herself unable to forgo them without being guilty of sin, he/she is to make a declaration to the praetor and the praetor would consult the senate... attended at least by 100 members. If the 100 members give permission, then he/she is allowed to perform the rite, but with not more than 5 people and without common fund of money, president of the ceremonies and no priest."

Livy goes on to refer to what happened after Postumius concluded his speech:

> "After the dismissal of the assembly the whole city was seized by extreme terror... which spread throughout the whole of Italy, as letters were received from friends telling of the decree of the senate."

> "The following night many people were caught escaping and were arrested..."

> "There was such a flight from the city that it resulted in a depopulation..."

> "Many names were reported to the authorities (...) some of these committed suicide."

> "Those who were found guilty of bacchic worship were condemned to death."

with the result that:

> "the people executed outnumbered those who were thrown into prison... All bacchic shrines were destroyed, except where an ancient statue or altar had been consecrated."

THE MOTIVES BEHIND THE REPRESSION AND ITS RESONANCE TODAY

The *Bacchanalia Affair* is a brutal yet revealing episode in Roman chronicles, which has been dubbed, for its extension and modality, the first major religious persecution in Western history (Gruen, 1990). However, if we give relevance, as is intended in this essay, to the fact that the rites of Bacchus were first and foremost an *ecstatic cult*, then we can affirm that the *Bacchanalia Affair* also brought about the first mass repression of people dedicated to the practice of altered states of consciousness in the West.

In a society that was increasingly dependent on the authorities' control over the minds and bodies of its citizens, a radical experience of freedom, communal excitement, bonding, power, joy, intoxication and excess like that of the *Bacchanalia* could not be tolerated for long. Such conviction, together with the predictable fear that the experience of loss of all control can excite over those who can't relate to its spiritual dimension, made the rites of Bacchus the perfect candidate for a ferocious campaign of moralisation. Moreover, by brutally repressing the *Bacchanalia* the Senate reached the convenient political objective of strengthening its authority in Rome and all over the Empire (Gruen, 1990)during a delicate phase of military expansion.

In other words the *Bacchantes* were killed so that a religious, political and moral order could prevail (Takács, 2000) and this was obtained at the expense of ancient practices that had accompanied Mediterranean civilizations from immemorial time in the experiencing of the sacred. The repression promoted the assumption that alteration of consciousness is not a viable means to come in contact with the divine; that ethical conduct is to be informed by the suppression of all bodily instincts and irrational behaviours and that social order is founded on the strict separation of classes organised in a hierarchy. Furthermore female leadership associated with secret nocturnal gatherings, the practice of ceremonial sexuality and the use of psychotropic substances was to be scorned and forbidden – a belief that was later reinforced by the *witch hunts*.

The *Bacchanalia Affair* registered the end of an era and the beginning of another, which has continued up to this day. And as we are now witnessing a surge of interest in the alteration of perception and its potentials, the *Bacchanalia Affair* can constitute a phenomenal precedent for a deeper understanding of both the sacred dimensions of consciousness-altering practices, and of the general attitude of authorities towards ecstasy.

Although Lucius Postumius seems to have originated a legacy of politicians who can still be heard echoing his own words and motives to repress practices of altered states, a new generation of 'devotees' might be now waking up Bacchus (and his mother) from their slumber. And in the endless succession of the cycles of history, a whole new phase in the practice of ecstasy, with a renewed awareness of its potentials, could be just about to begin.

Baldini, C. (2010). "Dionysus Returns: Contemporary Tuscan Trancers and Euripides' The Bacchae", chapter in Graham St. John (ed.) *The Local Scenes and Global Culture of Psytrance.* Routledge Studies in Ethnomusicology, pp. 170-185.

Baring, A. & Jules C. (1993). *The Myth of the Goddess: Evolution of an Image.* Arkana, Penguin Books.

Beard, M., North, J. and Price, S. (1998). *Religions of Rome. Vol. I. A History.* Cambridge University Press, Cambridge.

Burkert, W. (1987). *Ancient Mystery Cults.* Harvard University Press, Cambridge, MA.

Cassidy, W. (1991). Dionysos, Ecstasy and the Forbidden. *Historical Reflections.* Vol. 17, No. 1 pp. 23-44.

Coles, J. (2013). *How does Livy's Description of the Bacchanalia Scandal Reflect his Programmatic Aims?* Available: http://www.tsd.ac.uk/en/media/uniweb/content/documents/departments/lrc/thestudentresearcher/vol2no2journalarticles/JonColes_StudentResV2N2.pdf [2 November 2013].

Dashu, M. (2004). *The First Mass Hunt.* Available: http://www.suppressedhistories.net/secrethistory/romanhunt.html [16 August 2013].

Eisler, R. (1987). *The Chalice and the Blade.* Harper Collins, San Francisco.

Eliade, M. (2004). *Shamanism: Archaic Techniques of Ecstasy.* Princeton University Press, Princeton.

Gruen, E. (1990). *Studies in Greek Culture and Roman Policy.* University of California Press, Berkeley.

Harrison, J. E. (1991). *Prolegomena to the Study of Greek Religion.* Princeton University Press, Princeton.

Kerényi, K. (1992). *Dionysus: Archetypal Image of Indestructible life.* Adelphi Edizioni, Milano.

Limoges, S. (2009). Expansionism or Fear: The Underlying Reasons for the Bacchanalia Affair of 186 B.C. *Hirundo* 7: 77-92.

Meisner, D. (2008). *Livy and the Bacchanalia.* Available: http://www.scribd.com/ doc/246894101/Dwayne-Meisner-Livy-and-the-Bacchanalia#scribd [2 November 2013].

Meyer, M. (ed.) (1987). *The Ancient Mysteries: a Sourcebook of Sacred Texts.* Philadelphia: University of Pennsylvania Press.

Riedl, M. (2012). The Containment of Dionysus: Religion and Politics in the Bacchanalia Affair of 186 B.C.E. *International Political Anthropology.* Vol. 5 (2012) No. 2.

Streich, M. (2009) *Bacchanalia and Roman Repression: The Cult of Dionysus as a Threat to Republican Values.* Available: http://suite101.com/article/baachanalia-and-roman-repression-a109638_ [16 August 2013].

Takács, S. A. (2000) Politics and Religion in the Bacchanalian Affair of 186 B.C.E. *Harvard Studies in Classical Philology.* Vol. 100; 301-310. Harvard University Press. Cambridge, MA.

Walsh P. G. (1996) Making a Drama out of a Crisis: Livy on the Bacchanalia. *Greece & Rome.* Second Series, Vol. 43, No. 2, pp. 188-203. Cambridge University Press, Cambridge.

BRING WHAT YOU EXPECT TO FIND: PSYCHEDELICS & BRITISH FREE FESTIVALS

ANDY ROBERTS

Burning Man is a festival held in the high desert of Nevada that now attracts up to 50,000 people. The psychedelic experience is a massive driver for Burning Man as is the fact that those who attend need to be self-reliant. One goes to Burning Man as a participant not as a consumer and the event is lauded as being a revolutionary kind of festival. But history shows that Britain had exactly the same kind of event long before Burning Man was conceived (http://www.burningman.com).

These events were known as the Free Festivals. Their heyday was between 1970 and 1986, their names an invocation of Britain's psychedelic past: Glastonbury Fayre, the Windsor People's Free Festivals, Watchfield, Trentishoe, the Meigan Fayres, the Welsh Psilocybin festivals and more. These festivals operated under the same ethos as does Burning Man, one of self-reliance, and their motto was simply *Bring What You Expect To Find*. If you wanted something to be present, for example food, water, shelter, entertainment, crafts or drugs, you took responsibility to become part of the event by providing, to whatever degree, these things yourself.

My contention is that the British Free Festival movement was, like that of Burning Man, driven and underpinned by the use of and philosophy behind psychedelic drugs and it is possible to trace this idea by examining the roots of the free festival movement. And in doing so to recognise that it was the growth and popularity of these psychedelically inspired festivals and the culture which surrounded them which eventually led to Establishment efforts to dismantle that culture before its influence spread into mainstream society.

Free festivals were a response to a variety of emerging needs within the British counter culture. Night clubs and commercial festivals did not appeal to acid sensitised hippies who were questioning ideas of profit and control, and a demand arose for events generated by the counter culture itself, which would provide hippies with gatherings where they could celebrate, with like-minded people, their lifestyle. The spread of communes and the squatting movement in Britain's cities was also leading to a more communal way of life and it was a natural progression from communality in the cities to community in the countryside.

The annual Glastonbury Festival is a unique British cultural institution. Each year, on the weekend nearest the summer solstice, thousands of people flock to Worthy Farm in Somerset to camp for three days. Had it not been for the psychedelic focus of the first major Glastonbury event, the festival in its present form would not exist.

The 1971 event, known as Glastonbury Fair, was co-organised by Andrew Kerr, a researcher on Winston Churchill's biography and Arabella Churchill, Winston's granddaughter. On his way home from the commercial 1969 Isle of Wight rock festival Kerr, an LSD advocate, decided to stage a free festival, announcing to his friends,

> *"We've got to have a proper festival and it's got to have at least some cosmic significance. Let's have it at the summer solstice at Stonehenge."*　　　　　　　　　　　　　　*(Aubrey and Shearlaw, 2004)*

Kerr's intention to hold the festival at Stonehenge changed when he was given Michael Eavis' telephone number. A meeting was arranged and Kerr prepared by spending the night before meditating atop

Glastonbury Tor on LSD. The meeting was successful and, assisted by Arabella Churchill and utilising a small inheritance, Kerr formed Solstice Capers to organise the 1971 event.

Jeff Dexter, veteran London DJ, organised the music which consisted of the hippy bands that had played at the Roundhouse, including Quintessence, Brinsley Schwarz, Hawkwind, Gong, Traffic and Arthur Brown. These bands were open about their use of LSD and strove to create music and atmosphere to be enjoyed while under the influence of psychedelic drugs. Psychedelics of all kinds were freely available at Glastonbury, but LSD was prevalent. Williams Bloom's impression was that:

> "...nearly everyone was tripping at one stage or another. Sometimes it was being given away... The festivals would not have been what they were without hallucinogens."   *(Bloom, 2007)*

Arabella Churchill didn't indulge, but knew:

> "...there was a lot of acid because this man came up with a large briefcase and said: 'This is full of acid, man. I was going to sell it but everyone's doing everything for free so here, give it to everybody."
>   *(Churchill, 2006)*

The mixture of free psychedelics and living out the hippy ethos made the Glastonbury Fair the prototype for subsequent festivals. LSD brought people together at festivals, making the already otherworldly experience appear completely divorced from the Twentieth Century and Western civilisation. Mick Farren recognised:

> "We might as well have been in the sixth or even twenty-sixth century as we told tall travellers' tales of intoxication, of outwitting the law, of the lights in the sky, lost continents, the lies of government, collective triumphs and personal stupidity, while the music of past, present and future roared from the pyramid stage"   *(Farren, 2001)*

After the success of Glastonbury Fair moves were afoot to develop large free festivals that would be staged every summer. The Windsor, Watchfield and Stonehenge events, initiated and planned by key psychedelic visionaries, defined the free festival movement. Tracing the roots and motivations of these individuals demonstrates how crucial LSD was to the free festival movement and how the LSD experience underpinned the creation of environments in which people could live unhindered by what they saw as petty laws and restrictions.

The free festivals held in Windsor Great Park, near London, would not have happened without the efforts of one Bill 'Ubi' Dwyer, a crucial psychedelic mover and shaker. Dwyer became evangelical about LSD when living in Australia, where he was a major LSD dealer. LSD changed his life, "...it cleansed (me) of the evil of the past" (http://www.takver. com/history/aia/aia00034.htm), but his LSD dealing exploits led to him being deported. Finding himself in London he began dealing LSD again and also became involved with anarchist groups, helping produce *Anarchy* magazine's *Acid Issue.*

Dwyer was passionate about LSD and could often be seen in full oratorical flow at Hyde Park's Speakers Corner. His idea for a free festival came when he was tripping in Windsor Great Park where he had a vision of a "...giant festival in the grandest park in the kingdom." (McKay, 1996).

LSD was sold cheaply at the Windsor festivals and large quantities were given away, not always to a positive effect. Stage manager Roger Hutchinson recalls the 1973 festival:

> "...this chap came up with a brief case, quite smartly dressed, and said: 'Is there anyone I can talk to about the distribution of the contents of this. I've got all this acid that I want to give away'"

Hutchinson made an announcement over the PA,

> "Does anyone out there fancy getting a bit higher tonight, we've got some little tablets here, yours for the taking, free, and literally the audience just came up as one and went straight at the stage." (Hutchinson 2006)

Word spread among Britain's hippies that the 1974 festival to be held on the August Bank Holiday would be huge. The Windsor 1974 poster tantalisingly advertised the presence of 'psychedelicatessens', but although there were no psychedelicatessens, there was a great deal of potent LSD, the festival newsletter warning that the "brown acid is very strong, don't drop more than one tab." (Windsor Freep, August 1974).

Release set up a 'bad trip' tent and prepared for the inevitable acid casualties, one volunteer remembering:

"...the first wave of heavy trippers inundated us fairly early. An inordinate number of them seemed to be 'wankers' i.e. sexually repressed individuals liberated in a bizarre kind of way by the acid."

(http://www.ukrockfestivals.com/windsor-74-release.html)

Desire for LSD at free festivals was often outweighed by the effects of its potency and availability, Nigel Ayers remembering LSD being given away again at the 1974 event when he was

"...nearly crushed in the hand-out of hundreds of tabs of free acid"

(Ayres 2007)

Windsor veteran Allan Staithes recalls:

"There was a vast amount of acid at Windsor in 1974. Everyone was talking about it and it was obvious it was the focus of the festival. It was the strongest acid we had yet encountered and the afternoon dissolved in a blur of wild dancing and celebration."

(Staithes 2007)

The Release report noted:

"By late afternoon the area around the Release ambulance was reminiscent of a scene by Hieronymus Bosch... Worst of all were the juvenile philosophers who bellowed tedious cosmic observations about the state of the universe. (http://www.ukrockfestivals.com/windsor-74-release.html)

A notice board appeared at Windsor each day with suggested 'dropping times', the time at which LSD was best taken to enjoy the bands later in the evening. As the shadows lengthened, the weirdness began.

To enhance the LSD experience at free festivals several light shows had sprung up, projecting on screens behind the stages heated oil slides, films, and images of spirituality and nature. The most celebrated of these light shows was Acidica and as the name suggests, the Acidica crew were aware the drug of choice at free festivals was LSD. Acidica's John Andrews wrote, "We cater in the main for trippers."(http://festival-zone.0catch.com/acidia-lightshow.html).

Numerous religious sects were in evidence at most free festivals, offering friendship, advice, and free food but also ever-hopeful of gaining converts from the ranks of those left spiritually bereft by the LSD experience. The Divine Light Mission, followers of Guru Maharaj Ji, attended Windsor in 1974 where Premie Pat Conlon volunteered:

> *"I did service in the bad trip tent. I got stuck with one guy for most of one day. He was really freaked and two sisters in those ugly long dresses gave him satsang [devotional speech/chants etc.]. He ripped his clothes off and wanted to fuck them... He was masturbating furiously and unashamedly"*
> *(Conlon, 2007)*

Another Divine Light follower remembered:

> *"There were always at least 3 or 4 people being cared for at a time. I suppose I saw 50-60 people over the three days."*
> *(http:www.prem-rawt-talk.org/forum/posts/20130.html)*

An estimated 120 people had LSD experiences bad enough to warrant spending time in the Divine Light bad trip tent. Factor in similar organisations who dealt with bad trips and the individuals which were cared for by their friends and it's evident that hundreds of people had problems dealing with the strength of LSD on offer. Yet the potency of LSD at these events and the drug's power to rearrange personality and invoke spiritual experience only served to help create an alternative society to the materialistic culture of the 1970s.

During the same summer as the 1974 Windsor festival a smaller event was taking place at Stonehenge, the first in a series of free festivals there which continued until 1984. And, like the Windsor festivals, they were initially driven by one man and his LSD vision. 'Wally' was a cry heard echoing around festivals during the 1970s. Whether the original Wally was a lost dog or a lost soul, a hippy by the name of Phil Russell took Wally as his name, adding the surname Hope. Wally came from a wealthy background and he lived off a small trust fund; an early involvement with hippies during the 1960s embedded him firmly in the counter culture. But it was in Cyprus that the idea for a festival at Stonehenge formed. There, Wally had an LSD-inspired vision, seeing the sun as God and realising his mission was to reclaim the ancient sun temple of Stonehenge from the authorities to turn it into a place of psychedelic fuelled celebration and worship.

Wally had attended the 1973 Windsor free festival but was unhappy with the commercialism he saw there. The idea of hippies demanding money from hot dog salesman clashed with his belief in the concept of 'free'. Tim Abbott noted:

> *"He was disillusioned by what was already happening at the 'People's Free Festival'... He had a vision that it could be done in a purer way."* *(http://www.enablerpublications.co.uk/pdfs/notonly1.pdf)*

Wally and his friends set up camp at Stonehenge in the summer of 1974 in time for the solstice. Investigating officials from the Department of the Environment were told everyone there was called Wally, Wally's followers giving their names as 'Wally' to any enquirers such as journalists or court officials. For all Wally's tribal psychedelic dreams, the camp was a drab affair, with people sitting around doing very little, taking and discussing LSD and having debates about such ludicrous questions as whether children should be given the drug. Nor was this a theoretical discussion, one couple actually giving their children LSD on a daily basis, believing it to be the "...religious thing to do." (Guardian, 1974).

Wally returned to Cyprus for the winter of 1974 and on his return he threw himself headlong into the preparations for the 1975 Stonehenge festival. But it wasn't to be. In May 1975, at a squat in Amesbury, a police raid looking for an army deserter found Wally with LSD in his pocket. Bruce Garrard suggests the LSD didn't even belong to Wally but,

"He was arrested and brought before Amesbury magistrates where he not only didn't deny possession of LSD, he enthused about it, told them it was wonderful stuff which opened the mind to all kinds of visions..."

(http://www.enablerpublications.co.uk/pdfs/notonly1.pdf)

For the colourfully dressed Wally, with a reputation as a fervent anarchist and festival organiser, eulogising about LSD in a court of law was positively suicidal. The magistrates decided that he was mentally ill and detained him under the Mental Health Act.

Wally was held until two days after the 1975 Stonehenge festival had finished. LSD was Wally's sacrament, the chemical that drove his dreams, but the fire had been burned out of him with chemicals forced into him by the Establishment. He died a broken man in August 1975, choked on his own vomit. Stonehenge 1975 went ahead with up to 3,000 people in attendance and hundreds sitting in acid-drenched meditation to greet the summer solstice. Sid Rawle, the media styled 'King of the Hippies', had taken up the free festival cause and for the next decade was the prime mover in organising the Stonehenge events. Rawle had given up taking LSD, but was aware the majority of his constituents did and needed somewhere to celebrate and for many free festival goers psychedelics and spirituality went hand in hand.

Rawle also helped further link British psychedelic culture with an indigenous spirituality and many of the free festivals he organised were centred on prehistoric sites. He wrote to *The Times*, arguing:

"The evidence is indisputable that Stonehenge and the surrounding area is one of the most powerful spiritual areas in Europe. It is right that we should meekly stand in the presence of God, but it is proper that we should sing and dance and shout for joy and love and mercy that He shows us." (Rawle, 1978)

Not only was the counter culture psychedelically inspired, its fundamental belief system, a cosmic pantheism, was becoming firmly rooted and engendered in the psychedelic experience at ancient sites.

The rout by police at Windsor in 1974 meant there was no chance of another People's Free Festival being held there. But with the free festival movement now numbering tens of thousands, and the government fearing further violent clashes with police, Sid Rawle negotiated with Home Secretary Roy Jenkins, and the government offered an old airfield at Watchfield in Oxfordshire for a free festival.

The nine day Watchfield festival began on Saturday 23 August 1975 and once again the psychedelic experience underpinned and dominated the festival. Issue one of the Watchfield *Freek Press* pleaded on its front page: *"Serious acid shortage – send urgent messages out – the acid must get through"* (Watchfield Free Press, August 1975). This was an unequivocal statement from the festival organisers as to how important LSD was to free festival culture. And the LSD came through, in quantity and variety: blue blotter, blotter imprinted with the image of a strawberry, orange blotter and red and green microdot.

Of the other free festivals held in the 1970s, the Meigan and Trentishoe Fairs stand out as exemplars of the LSD-fuelled counter cultural lifestyle. The first three Trentishoe free festivals, ('72, '73 and '75) were held on the cliffs overlooking the Bristol Channel in North Devon where a small alternative community sprang up, offering free food, alternative energy and a variety of metaphysical groups. In 1975, 1,500 hippies grooved to the sounds of Hawkwind, Here & Now and the cream of Britain's free festival bands. Drugs were plentiful; vast amounts of cannabis had been donated to the festival by local dealers and LSD was freely available; Boss Goodman remembered, "Everyone was eating pink micro dots by the handful" (http://www.funtopia.pwp.blueyonder.co.uk/friends/bosstrentishoe.html). Bristol record store owner Nasher remembers "...the Chemists giving away bags of acid" (http://www.ukrockfestivals.com/trentishoe-73.html).

By 1976, the Trentishoe site had moved but LSD was still the preferred drug. In the words of one happy attendee:

"It was a hell of a show, everybody in the whole cell block (field) was playing with Lucy's Diamonds or Sam's Dice apparently. But I can still remember the utter madness of it now. It felt great."

(http://www.herenow.be/herenowpages/hn1976a.htm)

The Meigan Fair, held in the Preseli Mountains of West Wales, was also held over several years, attracting up to 8,000 people, the majority being hard-core members of the counter culture. In 1975, the Tipi people were out in force and the festival saw a range of live-in vehicles, foreshadowing the hippy convoys of later years. The press were favourably disposed to the festival, the event, one report noting:

"Naked she danced in the warm morning sun. Her hips swayed suggestively to the beat of the music. On her back was scrawled in ballpoint 'Got any Acid?'. On a rock nearby were chalked the words, 'Reality is an illusion caused by lack of LSD. Please, where can I score?'" *(Western Telegraph, 31 July 1975)*

The sheer amount of LSD available at the 1975 Meigan Fair has gone down in free festival legend. Chris Church remembers that on the first day of the festival:

"...almost all of them were incapable due to the huge consumption of psychedelics. It was the only event I ever went to where it seemed as if the lysergic state was the normal state to be in. Musically it was slightly chaotic..." *(http://www.ukrockfestivals.com/meigan-76-menu.html)*

So chaotic in fact that free festival synthesiser band Zorch were unable to play because one of their musicians, Basil Brooks, was "incapacitated by LSD" (http://www.ukrockfestivals.com/zorch.html).

Another Meigan veteran, Tim Rundall, remembered:

"...a very unhealthy quantity of liquid LSD which emanated, I believe, from the cottage that would later be raided by the Julie squad." *(http://www.herenow.be/herenowpages/vaultsyourtales2.htm)*

In the recollections of many free festival attendees there is a belief that much of the LSD available was made by Richard Kemp or Andy Munro, the chemists in the Operation Julie manufacturing and distribution conspiracy. This is possible. After his arrest Kemp claimed he had timed LSD production runs to coincide with the summer's festivals and he also claimed to have donated some of his profits to free festival organisers.

The Meigan Fair's organisers carefully considered the effects large and potent quantities of LSD would have and strove to create the best possible setting. Deep in the countryside the festival goers were untroubled by outsiders and there was always music, theatre or other events to occupy the minds of those who had indulged in psychedelics. For those who found themselves in the middle of an LSD-induced crisis there was a 'sanctuary' tent where trippers could be quiet, rest and read metaphysical literature.

If the prevalence of psychedelics at free festivals was kept as secret as possible from the outside world, the same cannot be said of the Welsh Psilocybin festivals. The clue was in the name! *Psilocybe semilanceata* had been used occasionally as a psychedelic by hippies since the late 1950s, but its use was limited until the mid-1970s. The sudden upsurge in usage was caused by a combination of factors; psilocybin mushrooms were free and freely available, they were organic and they echoed the usage of natural psychedelics by the indigenous tribal societies many hippy travellers emulated. Another reason for their sharp rise in popularity from 1976 onwards was the Operation Julie busts which led to a reduction in availability and strength of high quality LSD.

The origins of the Welsh Psilocybin festivals are murky but they appear to have begun in 1976, following the police eviction of the Elan Valley free festival. The event moved 16 miles up the road to a valley near Pontrhydygroes at Devil's Bridge which was at the centre of hills and moors awash with magic mushrooms. This was the first time a free festival had been directly and openly linked with the drug that generated its existence and word quickly spread about the event.

The *KOOL Newsheet*, published at the 1977 Meigan Fayre, referred to the 1976 mushroom festival held near Pontrhydygroes noting, "There was a big press coverage. Mushrooms seem to have caught the imagination of the 'public.'" (*KOOL Newsheet*, 1976).

By 1979 the festival had come of age and was widely referred to as the Psilocybin Fayre, attracting over 2,000 people from across Britain and the Continent. "Liberty 'trippers' flock to fungus festival" screamed the headline on the *Cambrian News*, and the hippies formally dubbed the site The Free State of Albion, while others gave it possibly more accurate names such as Crazy Creek (*Cambrian News*, 1979). When asked why he was attending the festival, one traveller replied he was there to "Worship Eris, the goddess of Discordia." (*Western Mail*, 1979).

One 1979 festival attendee was disappointed to find "There was no entertainment, just people and drugs"(Wood, 2014). This reinforced the notion that the Psilocybin festivals were primarily about the gathering and taking of magic mushrooms and the comin g together of a psychedelically motivated community rather than the consumerist entertainment prevalent at mainstream commercial festivals such as Reading, Knebworth and others.

Ceredigion District Council (CDC), already dismayed by the numbers of hippies settling in communes in the Aberystwyth area, decided that they wanted an end to psilocybin tourism and set about finding ways to prevent the festival taking place. An extraordinary meeting of the Council was called and a report entitled *Problems resulting from Hippie 'Festivals' in the Ystwyth River Valley* circulated. This contained some alarmingly prejudicial claims such as:

> *"The small scattered local population is quite at the mercy of a lawless, nameless and at the very least truculent and unsavoury army of occupation... It becomes noisy, filthy, repulsive – something between a fairground and a rubbish dump... We are polite because they seem slightly subnormal, and because we are afraid."* (Ceredigion District Council, 1980)

Various measures were mooted to disrupt the 1980 festival including flooding the site and asking local traders to discourage the hippies. But nothing was decided and the festival unfolded in the usual way. The consumption of psilocybin continued to underpin the festival's purpose, one festival goer recalling:

> *"The main event which usually happened 3 or 4 times daily was the arrival of the festival kettle – a blackened industrial sized affair, filled with magik mushrooms. The custom was to throw a few more shrooms inside, make as much tea as you all required, then carry it over to the next camp..."*
>
> *(http://www.ukrockfestivals.com/Psilocybin-Festival.html)*

As at most free festivals, many children were present and the *Daily Mirror* reported with horror that:

> *"...many took their children on a four-mile trek to gather dangerous 'liberty cap' mushrooms that give LSD-type hallucinations."*
>
> *(Daily Mail, 1980)*

Despite CDC's negative attitude, several journalists were less scathing. *The Observer* noted:

> *"Inside the festival there is a gentle carnival atmosphere. Food and mushrooms are shared... The festival is possibly the only meeting place of the hard core of 'new Gypsies' who travel with horse and cart and erect covered wagon-shaped tents, called 'benders', from bent over saplings... Many at the festival feel that the magic mushroom could lead to the return of the hippies as an important movement"*
>
> *(http://www.ukrockfestivals.com/Psilocybin-Festival.html)*

Free Festival veteran Bev Richardson was in attendance and reflected that:

> *"A lot of people come here to pick mushrooms to sell... The market for them isn't in this country but on the continent in Holland and Germany. I suppose you could call it a hippy cottage industry."*
>
> *(Richardson, 1980)*

Despite magic mushrooms being at least quasi-legal the police cracked down hard on the Psilocybin festivals. Festival Welfare Services' Penny Mellor commented that the atmosphere was spoiled by:

"...the high level of police activity on the approach roads to the festival. Road blocks were set up and virtually every person and vehicle going to the festival was stopped and searched during the weekend I attended. It was felt that this level of activity was sheer harassment of the festival people, on the pretence of searching for drugs, whereas the proportion of those actually found in possession of drugs was very small."

(http://www.ukrockfestivals.com/Psilocybin-Festival.html)

Mellor observed that although magic mushrooms may have been the event's purpose, this was no gathering of idle dole scroungers but was a valid representation of the alternative society which had grown from Britain's psychedelic culture:

"The Psilocybin Fair was a festival of doing. Almost everyone on site was trading in some way, mainly in food and crafts. People were very ingenious at thinking up new ways of exchanging money. The trading wasn't worked out on a high profit basis, but more on people working with whatever money and resources they had to generate enough money or basic supplies to live on themselves whilst providing a service for other people at the same time."

(http://www.ukrockfestivals.com/Psilocybin-Festival.html)

By 1982 the Welsh mushroom festivals were over, harassed out of existence by a combination of police and council tactics. But despite this harassment throughout Britain, free festivals survived the 1970s and thrived, going forward into the 1980s and becoming larger and even more drug focused. But the times were changing. Punk had happened and many festival goers were now driven by different agendas than community and drugs other than psychedelics. Free festivals were becoming less an environment in which to celebrate the psychedelic experience as an event at which Class A drugs could be freely traded, the threat of violence and rip off was never far away and theft from tents and vehicles endemic. The Stonehenge festival continued and despite the darkening days of the free festival, a survey done at the 1984 event showed just how strong the psychedelic allegiance still was. Of 500 festivalgoers who were drug users over 70% of these had used LSD and/ or Magic Mushrooms. Half of the users at Stonehenge intended to use LSD at the 1984 festival.

Events at Nostell Priory in 1984 and the trashing of the Stonehenge bound convoy in 1985 at the Battle of the Beanfield drew a bloody line under free festival culture. The dream was over. Small festivals still existed but were low key and rave culture now had its own events in the form of free parties and coalition between hippies and ravers at multi-day events such as 1992 Castlemorton rave. Many of the hard-core hippy travellers for whom the free festival circuit had become crucial to their lifestyle left for countries where they didn't feel persecuted. Others came off the road and pursued a house-based lifestyle.

In the second decade of the Twenty-First Century the psychedelic vision of community in the countryside lives on. Many small, albeit commercial festivals, such as Sunrise keep the spirit of those early free festivals alive and are still a suitable environment in which to enjoy psychedelics, and are still, to a certain extent, flag bearers for the ethos of Bring What You Expect to Find.

Aubrey C. & Shearlaw, J. (2004). Glastonbury: an oral history, Ebury Press, London, 2004.

Ayres, N. (2007). Personal Communication. 20 August 2007.

Bloom, W. (2007). Personal Communication by email, 9 August 2007.

Cambrian News (1970). *Liberty Trippers Flock to Fungus Festival. 14 September 1979.*

Ceredigion District Council (1980). *Problems Resulting from Hippie "Festivals" in the Ystwyth River Valley.* 25 February 1980.

Churchill, A. (2006). *Observer* music magazine. June 2006.

Conlon, P. (2007). Personal Communication. 18 August 2007.

Daily Mail, (1980). *Magic Mushrooms' hippies vanish.* 12 September 1980.

Farren, M. (2001). *Give the Anarchist a Cigarette.* Jonathan Cape, London.

Guardian Newspaper (1974). 26 November 1974. *Man gave LSD to son aged 9.*

http://festival-zone.0catch.com/acidia-lightshow.html.

http://www.burningman.com/.

http://www.enablerpublications.co.uk/pdfs/notonly1.pdf.

http://www.enablerpublications.co.uk/pdfs/notonly1.pdf.

http://www.funtopia.pwp.blueyonder.co.uk/friends/bosstrentishoe.html.

http://www.herenow.be/herenowpages/hn1976a.htm.

http://www.herenow.be/herenowpages/vaultsyourtales2.htm.

http://www.takver.com/history/aia/aia00034.htm.

http://www.ukrockfestivals.com/meigan-76-menu.html.

http://www.ukrockfestivals.com/Psilocybin-Festival.html.

http://www.ukrockfestivals.com/Psilocybin-Festival.html.

http://www.ukrockfestivals.com/Psilocybin-Festival.html.

http://www.ukrockfestivals.com/Psilocybin-Festival.html.

http://www.ukrockfestivals.com/trentishoe-73.html.

http://www.ukrockfestivals.com/windsor-74-release.html.

http://www.ukrockfestivals.com/zorch.html.

http:www.prem-rawt-talk.org/forum/posts/20130.html.

Hutchinson, R. (2006). Personal Communication. 20 October, 2006.

KOOL Newsheet. Meigan Fayre, 1976.

McKay, G. (1996). *Senseless Acts of Beauty.* Verso, London.

Rawle, S. *(1978). The Times.* 28 June, 1978.

Richardson, B. (1980). Observer, 14 September 1980, Festival of the Hippies.

Staithes, A. (2007). Personal Communication. 4 March, 2007.

Watchfield Free Press (1975). August, 1975.

Western Mail (1979). *Drugs quad kept busy at 'magic mushroom' festival.* 8 September, 1979.

Western Telegraph (1975). 31 July, 1975.

Windsor Freep (1974). August, 1974.

Wood, J.M. (2014). Personal Communication. 15 May, 2014.

THE MYSTICAL IS POLITICAL: FESTIVAL CROWDS, PEER HARM-REDUCTION AND THE SOCIOLOGY OF THE PSYCHEDELIC EXPERIENCE

DEIRDRE RUANE

Oh, is this the way they say the future's meant to feel
Or just 20,000 people standing in a field?
Pulp, 'Sorted for E's and Wizz'

Music festivals (especially those in electronic dance music) represent some of our culture's closest approaches to the liminal: in Victor Turner's work, an in-between state created through ritual practices in a consecrated space, perhaps reached via a pilgrimage or by crossing a symbolic threshold. People come to the liminal space to be transformed – to move from one state of being to another – and to experience unusual states of consciousness, including the sense of being apart from everyday space and time (Turner, 1969). Often this sense is boosted by psychedelic use. Psychedelic harm reduction organisations run by peers within festival subcultures provide sanctuary spaces, information and support with the aim that this might take place in the least damaging way possible.

However, the otherworld of the festival is not truly apart from the world and its laws. Peer harm reduction projects face challenges in their relations with festival organisers and law enforcement, within the dominant cultural narrative of prohibition and its attendant conspiracies of silence – for a first-hand account, see Ponté (2012).

The festivals counter the narratives of mainstream culture, to a greater or lesser extent, with their own professed politics and ideologies. These are often utopian in character, or rather heterotopian, creating a neutral space in which a patchwork of experimental ideas and ideals can jostle alongside each other (St John, 2001). Recurring themes include leftist politics and anarchism; environmentalism and sustainability; alternative spirituality; and a DIY ethos in which participation is preferred to being a spectator.

Thus politics and spirituality are often mingled, and inform each other, within the experience of a festival. However, what scholarship there is on the dance festival – apart from the body of work within the discipline of tourism studies and event management, which is beyond the scope of this paper – has tended to focus either on its political status or its place in modern-day practices of spirituality. In this paper I will examine ways in which these two themes intersect, with particular reference to the role of psychedelics and harm reduction in this process.

THEORY OF THE FESTIVAL

One body of theory on the festival with roots in anthropology, whose key proponent is Graham St John (2001), draws strongly on Turner and Van Gennep and focuses on its spiritual and ritual aspects, while also somewhat acknowledging its political side. In contrast, the other, springing from Birmingham-School-style cultural studies, seems largely unaware of the spirituality-focused work and instead concerns itself with the question of whether dance subcultures (and thus festivals in passing) have a distinct politics at all, generally concluding that they do not. In some formulations, their political expressions are said to be incoherent; Gilbert and Pearson (1999) point out that ravers have yet to approach the government with 'a clearly articulated list of basic political demands'. Alternatively, ideals are present but unstable: things of the moment which do not survive removal from the club or the festival site. Sarah Thornton (1994) cautions us not to confuse momentary feelings of dancefloor empowerment with 'substantive political rights and freedoms'. Finally, some theorists find no political meaning in dance culture at all. Reynolds (2012) describes the Thatcherist principles of consumption and entrepreneurship motivating early rave event promoters, suggesting that something fundamentally apolitical lies at its core.

Some early theories of ritual could be seen as support for this. Through its very isolation from the social structure the liminal space is said to reinforce the status quo, by making it even less imaginable to break the rules elsewhere or at other times. Meanwhile, while participants are in an open, accepting state of mind, the authorities take the opportunity to inculcate dominant cultural values (Turner, 1969).

However, liminality is also bound up with communitas, an experience of blissful mass unity in which participants feel at one with their community and by extension with all of humanity. Communitas is not inherently radical. In fact, Turner states that societies need it, and the social inversions and play of carnival, in order to remain stable (Turner, 1987). Nonetheless, communitas may take on a quality of radicalism in societies which no longer offer many legitimate opportunities for it.

Whether visiting and sharing supplies with neighbouring camps in Black Rock City, dancing to psytrance at a crowded outdoor stage, or taking part in open rituals like the fire ceremonies of Sunrise Celebration, many festivalgoers experience a communitarianism which is in short supply in the 'real world'. Gardner (2004) found bluegrass enthusiasts saw festivals as an escape from the isolation of their everyday lives. In festival crowds, people can experience 'collective effervescence', a concept originated by Durkheim to describe the effect of large-scale cultural rituals but applied by Michel Maffesoli to the exuberant sociality associated with being part of a fluid 'neo-tribe' connected by shared emotion (Maffesoli, 1996).

Many view this communal quality as specific to the festival time and space. However, it may contain the possibility of wider social change, especially when psychedelic experiences and the option of a sanctuary space to assist with them are added to the mix. This social impact is not dependent on the existence of an explicit political manifesto. It can take nothing more specific than a deeply felt and lived experience of being part of a collective to cause lasting change in the nature of one's connections with others and with society. However, the strength of this effect is modulated by the kind and circumstances of the experiences, and the extent to which we integrate them into our consciousness – a process in which psychedelic harm reduction can play a crucial role.

THE PSYCHEDELIC EXPERIENCE, INTEGRATION AND POLITICS

Disciplines running the gamut from neuroscience to anthropology are currently presenting new theories on how psychedelics function. Robin Carhart-Harris's studies of psilocybin using an MRI scanner (Carhart-Harris, 2013) conclude that the psychedelic state is one defined by greater suggestibility than normal consciousness, along with less sense of separateness between concepts or brain states (and thus, perhaps, between self and other). This can result in positive experiences involving 'magical thinking' and a benign, hopeful, perhaps utopian worldview, or negative, paranoid experiences that mimic psychosis – a clear demonstration of the importance of the right set and setting.

Other research by Carhart-Harris with David Nutt and the Beckley Foundation indicates that this suggestibility may have a mechanism similar to Aldous Huxley's 'reducing valve' theory about the effects of mescaline in *The Doors of Perception* (Huxley, 1954): decreased blood flow to areas of the brain acting as 'connector hubs' seemed to suppress their usual 'censoring activity', allowing for new patterns of thought (Feilding, 2013).

With reduced access to our everyday schema of the world which filter sensory input based on past experience, new models of self and other may have the opportunity to develop. Echenhofer's research on ayahuasca drinkers resulted in his 'creative cycle processes model': a common pattern in the ayahuasca experience in which a dissolution of the self and its assumptions (often traumatic) is experienced, followed by the creation of new self-concepts and/or views of the world, concluding with the integration of these new models into the self as a whole ('vertical integration') and then into the subject's concept of the world ('lateral integration') (Echenhofer, 2012).

We must be wary of what Letcher (2013) calls the "common core model" of mysticism, a tendency to focus on common cross-cultural (and cross-substance) themes at the expense of awareness of cultural diversity and the sheer bewildering variety of psychedelic experience. That said, Echenhofer's study, the psilocybin research of Griffiths et al. (2006, 2008), and many first-hand accounts do seem to exhibit a common thread: a trajectory towards a sense of interconnectedness or consolidation, especially if the experience is allowed to run its course in a supportive setting. However, a similarly prominent theme is a period of dissolution, fragmentation or isolation early in the experience, corresponding to Echenhofer's stage of schema breakdown – a period which can be painful or difficult.

I contend that working through, or bypassing entirely, this isolating stage and arriving at the stage of integration and interconnectedness – with or without assistance – can be transformative not only personally but politically. This is because such a state of consciousness runs dramatically counter to cultural narratives deeply embedded in, and helping to perpetuate, Western neoliberal society.

NEOLIBERALISM AND THE FEAR OF CROWDS

After 21 people were suffocated in a crowd at the Duisburg Love Parade in 2010, Luis-Manuel Garcia (2011) studied the online reaction to the disaster. He wrote:

> *"Numerous commentators on the web… seemed to be coming to the same conclusion: there's something intrinsically wrong with large crowds, and by extension there's something wrong with people who are drawn to them."*
> *(Garcia, 2011)*

The commentators in his analysis constantly invoked the loss of the rational self and the return to a bestial nature thought to be inherent in participation in a crowd, using phrases like 'lizard brain', 'herd animal', 'knee-jerk', and so on. People who voluntarily engaged with large crowds were looked at suspiciously; as another of the web commenters put it, "You know, you could have the same dancing, love, drugs, and whatever in the comfort of your own home with friends" (ibid.)

The comments on a recent pro-communitarian article by Giles Fraser in the Guardian (Fraser, 2013) have a similar tone, identifying communitarianism variously with terrorism, fascism, Soviet gulags and Stalin's purges. Not only the 'herds' and 'mobs' of festival crowds but any large groups of people are portrayed as inherently dangerous.

This currently pervasive attitude, suggesting that people need to be insulated from each other for fear of a conflagration, is a relatively recent invention. According to Garcia (2011), the pathologising view of crowds originated in late 1800s sociology, but was immediately embraced by opponents of universal suffrage ("why leave the nation in the hands of an 'electoral mob' when an elite aristocracy could handle it with cool professionalism?").

The neoliberal elites of today also find it convenient. Garcia quotes Mazzarella: crowds are 'the past of the (neo)liberal democracies of the global North', a past they wish to distance themselves from (ibid.). A salient feature of neoliberalism is its tendency to fragment communities and isolate individuals. Purcell et al. (2010) document Thatcher's efforts to create a 'mobile workforce' by weakening community and extended family ties so that individuals would be free to move to wherever jobs happened to be. Gill (2009) shows how marketisation and emphasis on competition among academics has undermined solidarity and isolated them from each other, while Ehrenreich (2005) describes how neoliberal attitudes to self-actualisation in the job market serve elites by masking wider social inequalities.

Whatever it stems from and whoever it really serves, fear of collective experience is deeply embedded in our current cultural narrative. Terry Wassall (2010) writes of neoliberalism that it so permeates all aspects of our lives as to make alternatives unimaginable. Left-wing thinkers' proposed solutions are still founded on neoliberal assumptions about the meanings of freedom, choice, commodities and markets. Similarly, fear of crowds appears as common sense: avoid them or risk losing your mind.

THE POSITIVE CROWD EXPERIENCE

In all of this condemnation, the idea that there are positive kinds of crowd experience is rarely put forward, though festivalgoers, free partiers and clubbers experience them on a regular basis. However, Garcia (2011) describes a benign sort of ego dissolution:

> *"One of the possible pleasures of partying in a crowd is the sensation of coming undone, of feeling your sense of a bounded, unitary self unravel and fray at the edges. For some people, this sense of being temporarily relieved from the compulsion to be a consistent, coherent subject can feel like total bliss."*

This, he writes, cannot be achieved in small groups:

> *"Part of the experience of being in a crowd [is] about the sheer thrill of feeling one's affects and emotions being mirrored, amplified, and circulated a thousandfold"*

The mindless automaton figure described by the pathologising view of crowds seems to be an unhelpful caricature. All the same, it would appear that to some extent being in a crowd does constitute an altered state of consciousness in its own right. The quote above about the dissolution of self and the mirroring of mood and gesture parallels Carhart-Harris's description of the psychedelic state (Carhart-Harris, 2013). Even in the admittedly unlikely event of unanimous sobriety, festive crowds are liable to be high in openness and suggestibility, and low on ego boundaries. This invites speculation as to how the effect might be boosted if many in the crowd were also on psychedelics.

One of my interview participants in a small study on extraordinary and spiritual experience at festivals (Ruane, 2011) gave me this account of a unitive experience:

> *"And I'm sat on the edge of the crowd at the back of the tent, and I could feel all this energy going through me and I could see all these – webs of light everywhere… like, where everyone was and where they'd been and where they were going… just this feeling that everything really is connected and everything really is made of energy."* (KP)

Though she had taken a small dose of a psychedelic the previous day, KP stated she was not under the influence when this occurred. Instead, her account foregrounds the effect of being in an ecstatic crowd: the collective experience itself as consciousness alterant.

THE ROLE OF HARM REDUCTION

Festivals are for many the best available setting for such experiences, but they can also be chaotic, unpredictable, and prey to the whims of weather, along with the risks attendant on becoming disinhibited and vulnerable among strangers, not all of whom have good intentions. Under these circumstances, the consolidation phase of the psychedelic experience may be difficult to reach. Enter harm reduction services, which aim to provide at least one place on site where the environment is favourable.

In one welfare worker's experience, general awareness that there is a sanctuary tent on site leads to fewer people having the sort of problems which might cause them to need one:

"...we're a safety net. If you're teaching somebody how to do circus skills and they're learning how to walk a tightrope, you put a safety net underneath, and that gives people the confidence to walk across that rope without ever falling into the safety net" (Bill)

We now consider those who do make use of the safety net. The peer harm reduction group Kosmicare UK has kindly allowed me access to five years of their records. They contain many examples of psychedelic experiences moving from isolation to integration under their supervision: from those of mythic, archetypal proportions (for example: a woman who said she had 'lost the light'; a man who believed he was in hell and worked through the experience with the facilitators until they arrived at a different narrative of reliving and releasing past traumas) to more simple and practical examples of disconnection (a young man who was convinced he had wandered into a completely different festival). Kosmicare UK visitors are sometimes initially unwilling to come to the service for help due to a sense of 'not belonging'. In many of the cases this is said to transmute into integration and belonging – most strikingly in the case of a young woman, at first afraid to enter the space, who ended up acting out a psychodrama in which she asked the two facilitators to stand in for her parents so that she could begin to voice long-repressed issues.

Another account from my extraordinary experience interview project (Ruane, 2011) shows the isolation-integration trajectory from the point of view of a harm reduction service user. The participant was alone at a large festival in Europe and had taken LSD:

"I saw the whole majesty of the universe, just laid out above me, and the tiniest tiny specks saying 'You are here, and you're all alone!'... And so I retreated [to the sanctuary tent... that was like having a year's worth of psychotherapy in one night, it was exactly what I needed... I didn't feel like I fitted into the space of the festival as a

whole, or like I belonged there. Whereas at that moment, I very much fitted into the purpose of the sanctuary tent, and therefore – that's where I went." (DQ)

So what are the long-term results of a successfully integrated psychedelic experience? Griffiths' follow-up study on his psilocybin participants (Griffiths, 2008) found them to score higher than initial pre-study tests on a battery of measures of well-being 14 months after the initial study. MacLean et al. found long-term change in the fundamental personality trait of openness, a depth of change almost unknown in adulthood (MacLean, 2011). Tramacchi (2004) found that in many cultures psychedelic rituals seemed to lead both to an enhanced sense of self and to stronger bonds with the community. Transpersonal therapist Renn Butler says of those who have undergone a full mystical experience that they exhibit "critical attitudes towards the abuse of power... higher responsibility, positive ethics, and utmost respect for life" (Butler, 2009).

Butler implies a common ideology inherent to the psychedelic experience. Other commentators might disagree, finding such experiences to be wildly varied or simply indescribable. However, integrated unitive experiences can be seen as inherently political acts merely by virtue of running counter to, and suggesting alternatives to, the all-consuming dogma of isolation – however ineffable or resistant to analysis the core of the experience may be. Furthermore, awareness of interconnectedness is political in that it encourages thought about the systems in which one is embedded and the power dynamics at play within them.

In addition, the recent research on psychedelics discussed above suggests that, by loosening the hold of schema on our thinking, they may help solve the problem previously pointed out by Wassall (2010) by making it easier to think the normally unthinkable. Adams describes how use of psychedelics can promote an understanding of processes of schismogenesis – the rifts that develop between society and its 'others', for example disaffected youth (Adams, 2014). This effect is far from guaranteed; Riley et al. (2010) showed how neoliberal philosophies persisted, competing with discourses of connectedness, in the talk of a group of magic mushroom users. However, at the very least, experiences of communal festivity while in a state of psychedelic openness offer a chance to perceive and criticise forms of social disconnection normally taken for granted.

CONCLUSION

The spiritual and political approaches to festival theory, far from being separate, can be viewed as intricately interwoven. Positive experiences both *among* crowds and *of* crowds, at a time when the mind is amenable to new schema, have the potential for lasting change not just of the individual but of how they relate to others and perceive their place in the social world – in a way which may not be achievable in a therapist's office or in the comfort of one's own home.

Some of the current discourse on the benefits of psychedelics seeks to draw a bright line between therapeutic/clinical and recreational use, and to distance itself from recreational use, presenting it as damaging and chaotic. However, aside from the fact that in the absence of regulated use, recreational settings are for many the only point of access to the therapeutic modalities of these substances, experiences in recreational settings have their own value and social significance. They are not cleanly separable from therapeutic use, and othering these users and their practices in search of legitimacy would be regrettable.

In the meantime, let us return to the stated remit of harm reduction: to reduce the risk of drug-taking behaviours. Accordingly, it behoves us to provide support through which as many of these experiences as possible can safely run their course rather than being arrested at the fragmentation/isolation phase. The emergent effects on society might surprise us.

Adams, C. (2014). Psychedelics and shadows of society. *openDemocracy*. Available at: http://www.opendemocracy.net/cameron-adams/psychedelics-and-shadows-of-society-0 [Accessed February 27, 2014].

Butler, R. (2009). *Archetypal Astrology and Transpersonal Psychology: The Research of Richard*

Tarnas and Stanislav Grof. Available at: http://www.docshut.com/xnnvr/renn-butler-archetypal-astrology-and-transpersonal-psychology-the-research-of-richard-tarnas-and-stanislav-grof. html [Accessed February 27, 2014].

Carhart-Harris, R. (2013). Psychedelic Drugs, Magical Thinking and Psychosis. *Journal of Neurology, Neurosurgery and Psychiatry*, 84(1).

Echenhofer, F. (2012). The Creative Cycle Processes Model of Spontaneous Imagery Narratives Applied to the Ayahuasca Shamanic Journey. *Anthropology of Consciousness*, 23(1), 60–86.

Ehrenreich, B. (2005). *Bait and Switch: The (Futile) Pursuit of the American Dream.* Metropolitan Books.

Feilding, A. (2013). Breaking conventions in policy and science. C. Adams et al. (eds.) *Breaking Convention: Essays on psychedelic consciousness.* Strange Attractor, London.

Fraser, G., (2013). The west is in thrall to Kantian ideals of personal freedom. And suffers for it. *Comment is free | The Guardian.* Available at: http://www.theguardian.com/commentisfree/belief/2013/sep/20/west-kantian-ideal-freedom-society-suffers [Accessed February 27, 2014].

Garcia, L. M.. (2011). Pathological Crowds: Affect and Danger in Response to the Love Parade Disaster at Duisburg. *Dancecult: Journal of Electronic Dance Music Culture*, 2(11).

Gardner, R. O. (2004). The Portable Community: Mobility and Modernization in Bluegrass Festival Life. *Symbolic Interaction*, 27(2).

Gilbert, J. & Pearson, E. (1999). *Discographies: Dance, Music, Culture and the Politics of Sound.* Routledge, London.

Gill, R. (2009). Breaking the silence: The hidden injuries of neo-liberal academia. In R. Gill & R. Flood, eds. *Secrecy and Silence in the Research Process: Feminist Reflections.* Routledge, London.

Griffiths, R. et al. (2006). Psilocybin can occasion mystical-type experiences having substantial and sustained personal meaning and spiritual significance. *Psychopharmacology*, 187; 268–283.

Griffiths, R. et al. (2008). Mystical-type experiences occasioned by psilocybin mediate the attribution of personal meaning and spiritual significance 14 months later. *Journal of Psychopharmacology*, 22(6); 621–632.

Huxley, A.. (1954). *The Doors of Perception.* Chatto and Windus, London.

Letcher, A. (2013). Deceptive cadences: a hermeneutic approach to the problem of meaning and psychedelic experience. C. Adams et al. (eds.) *Breaking Convention: Essays on psychedelic consciousness.* Strange Attractor, London.

MacLean, K., Johnson, M. & Griffiths, R. (2011). Mystical experiences occasioned by the hallucinogen psilocybin lead to increases in the personality domain of openness. *Journal of Psychopharmacology*, 25(11); 1453–1461.

Maffesoli, M.. (1996). *The Time of the Tribes: The decline of individualism in mass society.* Sage, London.

Ponté, L. (2012). Zendo Project 2012: Harm Reduction in the Black Rock Desert. *MAPS Bulletin Annual Report.* MAPS.

Purcell, C., Flynn, M. & Ayudhya, U. C. N. (2011). The Effects of Flexibilization on Social Divisions and Career Trajectories in the UK Labour Market. H. P. Blassfeld et al. (eds.) *Globalized Labour Markets and Social Inequality in Europe.* Palgrave Macmillan. Available at: http://www.palgraveconnect.com/pc/doifinder/10.1057/9780230319882 [Accessed April 9, 2014].

Reynolds, S. (2012). *Energy Flash: A Journey Through Rave Music and Dance Culture*. Soft Skull Press, Berkeley.

Riley, S., Thompson, J. & Griffin, C. (2010). Turn on, tune in, but don't drop out: The impact of neo-liberalism on magic mushroom users' (in)ability to imagine collectivist social worlds. *International Journal of Drug Policy*, 21; 445–451.

Ruane, D. (2011). 'Time was away and somewhere else': the music festival as a context for extraordinary experience. Conference presentation, 23 September 2011, *Exploring the Extraordinary*. University of York.

St John, G. (2001). Alternative Cultural Heterotopia and the Liminoid Body: Beyond Turner at ConFest. *Australian Journal of Anthropology*, 12(1); 47–66.

Thornton, S. (1994). Moral Panic, the Media and British Rave Culture. Ross, A. & Rose, T. (eds.) *Microphone Fiends: youth music and youth culture*. Routledge, London.

Tramacchi, D. (2004). Entheogenic dance ecstasis: cross-cultural contexts. St John, G. (ed.) *Rave Culture and Religion*. Routledge, London.

Turner, V. (1969). *The Ritual Process: Structure and Anti-Structure*. Routledge, London.

Turner, V. (1987). Carnival, Ritual and Play in Rio de Janeiro. In Falassi, ed. *Time out of time: Essays on the festival*. University of New Mexico Press, Albuquerque.

Wassall, T. (2010). Anti-capitalism: thinking the unthinkable? *The Sociological Imagination*. Available at: http://sociologicalimagination.org/archives/777 [Accessed February 27, 2014].

REMIXTICISM: DMT, PSYCHEDELIC ELECTRONICA AND MYSTICAL EXPERIENCE

GRAHAM ST JOHN

This article illustrates how the sonic mythography and visionary artistry of our times have been cultivated with the assistance of DMT (N,N-Dimethyltryptamine). I focus specifically on the influence of DMT on psychedelic electronic music and culture. I pay attention to the gift Terence McKenna bestowed upon key figures in the early Goa Trance development, which inspired a transnational techno-mystical movement with distinct music aesthetics and event-culture. I explore how DMT consciousness has infiltrated psychedelic trance, where the *'sampledelic'* artifice of electronic music production and the sampling of entheogens afford mystical experience – what I call *remixticism*. The works of Shpongle, Alex Grey and others are discussed as multi-mediations of the gnostic payload of hyperspatial transit.

In the arts of remixticism, media and molecules are adapted to the purpose of consciousness: its dissolution, its expansion and its evolution. In psychedelic electronica – and specifically psychedelic trance (and before it Goa Trance) – digital and chemical elements are sampled and remixed to the goal of entering extraordinary, visionary, mystical states of consciousness. In the global psychedelic festival culture that has evolved over the last 20 years since Goa Trance, if we trace the lines of this remixticism, we find ourselves inside the "vibe" – the highly optimised socio-sonic aesthetic – that is the raison d'etre of this culture. I am interested in how DMT became an important agent in this development.

This is a story with which I became familiar while researching for my recent book *Global Tribe: Technology, Spirituality and Psytrance* (St John, 2012), where I journey into the heart of psyculture and its percolating vibe. That book meditates on spiritual technologies that are implicit to the global profusion of psychedelic trance, especially the festivals mounted downstream from Goa Trance. It addresses techniques employed by music producers to simulate and stimulate the experience that is mystical, transcendent and liminal. Among those techniques is the DJ-producer's practice of sampling vocal fragments from a plethora of sources (e.g. film, TV, spokespeople), programmed in tracks to evoke altered conditions of mind and body. These fragments are what I call nanomedia. In psychedelic electronica, media shamans sample from cinematic sources, the commentaries of altered statesmen, and the narratives of unpopular culture to fashion revelatory soundscapes infused with a gnostic sensibility.

> *"'Do you want to know what it is?' asks Captain Morpheus from The Matrix. 'It's all around us, here even in this room. You can see it out your window, or on your television. You feel it when you go to work, or go to church or pay your taxes. It is the world that has been pulled over your eyes to blind you from the truth.'"*

As electronic remixologists and data-miners hack the scripts of movies, computer games and celebrities, whose narratives are edited, repurposed and fused to the prologues and breakdowns of tracks mixed by DJs for interactive receiver cults on dancefloors worldwide, culture is rescripted to esoteric ends.

"We're not dropping out here, we're infiltrating and taking over." This is Terence McKenna sampled on Eat Static's 'Prana' (*Abduction*, 1993), among the first tracks to amplify the warm meandering lilt of a figure who would become the nanomediated voice of psyculture. A bard-like altered statesman, McKenna celebrated DMT to be among the most powerful vehicles for hyper-dimensional transit. "I remember the very very first time I smoooooked DMT," he reminisces on 1200 Micrograms' 'DMT' (TIP.World, 2002). The words are from a Parallel Youniversity lecture at London's Megatripolis in 1993, evoking a life-changing event in Berkeley in 1966/67. Having enrolled in the short-lived Experimental College Program at UC Berkeley, Terence announced to his brother Dennis at that time, as reported in Dennis' *The Brotherhood of the Screaming Abyss*, that "I know what the philosopher's stone is; it's sitting in that jar right there on the bookshelf" (2012). With heavy doses of science fiction and Jung, the McKennas regarded DMT as "the key" to the door of "the Secret," an understanding that explains their subsequent efforts, notably the February 1971 experiment at La Chorrera in the Columbian Putumayo, Amazonias, recounted in Terence's entheographic tour de force *True Hallucinations* (T. McKenna 1993), and a prolific speaking schedule the voluminous contents of which are now uploaded into eternity, animating the strange adventures of the "machine elves of hyperspace" among occupants of alternative households worldwide. Since the 1990s, references to DMT have multiplied in psytrance productions – in which McKenna posthumously remains the most commonly sampled individual. Materializing like a spectre in a séance, he will sometimes offer instructions from the beyond as he does on the title track of Cosma Shiva's (aka Avihen Livne and Jörg Kessler) *In Memory Of Terence McKenna* (2001): "...vaporize it in a small glass pipe". Like a familiar speaking on behalf of the multitudes whom continue to break into worlds recognised as hyperspatial dimensions, his voice and ideas are immortalised in the music. For instance, on 'Power and Light' by Streamers (*Power and Light*, 2012) an announcement from an invisible podium vibrates across open air stomping grounds:

"We have no idea what it would mean in our own lives if we could throw off the notion of ourselves as fallen beings. We are not fallen beings. When you take into your life the gnosis of the light-filled vegetable... the first thing that comes to you is: you are a divine being. You matter. You count. You come from realms of unimaginable power and light, and you will return to those realms".

Psychedelic trance has quite literally been smudged with DMT. It would be difficult to locate an earlier DMT-influenced project than Shpongle, whose co-founders Raja Ram and Simon Posford were directly inspired by DMT and McKenna. The landmark album *Are You Shpongled?* (1998) had sung the ode to DMT (and McKenna). The midsection of the album track 'Divine Moments of Truth' may be as recognisable within the psychedelic diaspora as the Coca-Cola iconography is to the rest of the world: "DMT, DeeMT, DMT, DeeMT LSD do DMT, LSD do DMT." That album also features compositions that are among the first attempts to convey the DMT experience using samples from popular cultural sources.

"It's three o'clock on what may well be the most important afternoon of the history of this world – humanity's first contact with an extra-terrestrial species."

The line is on 'Vapour Rumours', a track on *Are You Shpongled?* also featuring:

"...we're receiving the transmission. We're seeing some sort of vapour. I don't know, some sort of gas or something. Wait, something's happening."

In the reprogramming of these samples, the second stripped from an episode of the 1995 revival of the TV series *The Outer Limits*, the extraterrestrial alien contact gnosis so prevalent within Space Age psychedelia is repurposed for the Hyperspace Age. The vapour drifting through this dialogue and mooted in the track title is a thinly veiled reference to what is produced when DMT is smoked – vapourised – from a freebase chillum. In the reprogramming of the sample, the extraterrestrial 'contac' narrative is remixed and repurposed to evoke the interdimensional entity contact so frequently reported among DMT users (Luke, 2011).

While few will have recognised the meaning behind those samples in 1998, over a decade later at Portugal's Boom Festival, the vapours had escaped from the pipe, and DMT was no longer a rumour. In the Dance Temple at Boom 2010, the following lines washed over me and thousands of fellow worshippers as DJ Treavor (Walton) from LA's Moontribe, performing within a stellated dodecahedron DJ portal, unleashed the zeitgeist care of Loud's remix of Quantize's "Dymethyltryptamine" (sic) (Sphera – *Wide Open*, 2011): ".....heavy doses of Dimethyltryptamine." It was a superliminal suggestion, echoing and amplifying the desirable altered state, quantised into the rhythm as the vapour of a freshly smoked spice blend hangs pungent in the air.

Nanomedia is programmed for global dancefloors in tracks that serve as sonic billboards. If we listen to the sound fragments broadcast in a dizzying diablo of productions, it's like tuning in to an underground radio show, where the news is delivered in cut up. On the cutting-room floor, event-goers are exposed to important announcements. Thus Alex Grey reports:

"DMT is one of the most potent hallucinogens known to humanity. Most often it's smoked, and when you do that you break through very rapidly within 30 seconds to, say, a peak of an LSD experience and so alternative realities come up very fast".

Onboard 'Baraka', included on their 2006 self-titled debut album, F.F.T. (aka Future Frequency Technology) transmit the news in brief which is repeated in the breakdown:

"...for the vast majority of people, DMT was the most powerful psychedelic experience of their entire lives."

In other news, author of *Supernatural: Meetings With the Ancient Teachers of Mankind*, Graham Hancock does the journaling:

> *"DMT is astonishingly widely available in plants and animals all around the world and so far nobody really knows why its there or what its function is."*

And again, Hancock reviewing two key possibilities in condensed and digestible form transcribed for participants of the vibe on the track 'Imprint' by Ritmo and Zen Mechanics (Ritmo, *Phrase A*, 2011):

> *"We are dealing with free-standing parallel dimensions which our consciousness is able to enter in these states. But another possibility that has to be looked at, is that this information, the teaching program, that is unleashed by these plants may in someway be imprinted on our DNA."*

With these select editorials and cuttings, psytrance productions programme the news, while at the same time acting as media of transmission for DMT consciousness.

But let me further trace the background of this story. Shpongle's single *Divine Moments of Truth* (2000) features the track 'Shpongle Trance Remix' on which Raj divulges his initial journey into DMT-space: "it was like a gigantic creature, which kept changing shape", motioning towards his intention near the end of the track: "I want to try it again real soon, and take more." In conversation with Raj prior to his 70th birthday in 2010, he confirmed to me that Shpongle had been inspired by DMT, and specifically by a session he had with Simon Posford where a giant octopus "put its tentacles into my third eye and pumped the information directly into my brain" (Raj Nov 13 2010). In the mid-1990s:

> *"a new world full of entities and spirits that were benign.... guided me... spoke to me... took me... to Shpongle land... and it was directly because of the DMT... that Shpongle and the word Shpongle was given to me through channelling and vision."*

As vocal samples on Shpongle's debut and subsequent albums attest,

DMT had a prodigious impact on Raj. And Shpongle was commissioned as its emissary class Starship. Debut release *Are You Shpongled?* – a reference to Hendrix's *Are You Experienced* (1967) – is infused with the elf spice. The word 'Shpongle', and the experience of 'enshponglement', have become a conspiratorial nodding and winking experience for trance travellers and psychonauts worldwide.

But while most of this information was not new to me, my eyebrows lifted when Raj told me that a packet containing three to four grams of pink DMT crystal had landed in his hands in 1992/93. He referred to this as the "pink packet". Raj claims he introduced 100 people to the contents of this divine stash bag mostly on the beach or in his house in Anjuna, Goa, where people "were transported into a fabulous new realm". "DIVINE MOMENTS of TRUTH" was painted across the door arch, and every pillar and wall in the house was painted with DMT inspired art. What he referred to as the pink packet caused me to recall Michael Hollingshead's infamous mayonnaise jar full of LSD paste – from which Leary and innumerable others were dosed in the early 1960s (Stevens 1987).

But the story goes back further than this, to English DJ and promoter Nik Sequenci, who played ambient music to back Terence McKenna's UK lecture tours in the early 1990s, and who with the help of McKenna received 11 grams of pink DMT crystal. About his first DMT revelation, he stated:

> *"I had elves and fairies playing music to me, which was psychedelic trance, and they were saying 'go back to Earth and make this music.'"*

He needed artists capable of translating this revelation into music. I've already mentioned Raja Ram, a key translator who seemed well chosen given that he represented a bridge from the 1960ss to the 1990s. Indeed Raj would champion and cultivate DMT consciousness in a fashion not unlike the embrace of LSD in the earlier psychedelia. Experienced with acid in late 1950s Greenwich Village, becoming flautist for the psychedelic jazz outfit *Quintessence*, Raj was turned on all over again 35 years later. Only this time, as he wrote to me,

"Even acid never went that far, that internal, that BIZARRE AND STRANGE... DMT... it was the most amazing experience ever."

But before Sequenci met Raj, he made an important pilgrimage to Goa in late 1991, where he met Martin 'Youth' Glover. It was like love at first sight: "I had an inner smile cause I knew he was the man". Referring to the man who founded the first Goa Trance label, Dragonfly Records in 1993, Sequenci said that Glover "turned his hand to electronic music after his DMT experience, and psychedelic trance was born." Moreover, when that music was released, he heard "the music that the elves and fairies and pixies played." It was "the most moving experience" for Sequenci, because he'd found the man who could translate the sounds of hyperspace.

Sequenci eventually hand-picked about 120 people to receive the pink DMT. Along with Glover and Raja Ram, among the chosen few were Mr C (from The Shamen), Simon Posford and Space Tribe, all of whom first smoked DMT between 1992–95. These were germination years for Goa Trance, for which Sequenci is an unsung cultural midwife. Originally sound engineer at Dragonfly, and later founder of Twisted Records, and co-founder of Shpongle, Simon Posford - among the most respected artists in the psychedelic diaspora - is unequivocal about the impact of DMT on his career. Interviewed by Alex Grey in the first edition of *CoSM* magazine, Posford referred to his first hit – from Raj's pink packet – confirming that:

"DMT provided me with the most inspirational, life-changing, religious, mystical experience I've ever had... You could do yoga and meditation for a hundred years and maybe not get to the place you get to with thirty seconds on a pipe. Meeting some kind of über-consciousness or entity during one DMT experience, I could see the thoughts from my brain as visual hieroglyphic, holographic multidimensional hallucination, flowing to this entity that was getting off on them. I heard a little flurry of flute from music that Raj had created and which I had to INSIST that he create in the studio to use in a piece of music."   *(in Grey, 2005)*

The *Are You Shpongled?* track 'Monster Hit' evokes a gargantuan alteration of consciousness through exotic soundscapes, subterranean caverns and angelic voices, with wayfarers traveling upon the notes blown by Raj. In an interview with David Jay Brown (2011), Posford confirmed that it is 'Behind Closed Eyelids' which features an imitation of the divine riff that "we had been instructed to use by the alien creatures we encountered on DMT". None other than Aldous Huxley is recruited to announce:

> *"...behind closed eyelids, in very many cases, the visionary quality, the quality of the visions so to say spills over into the external world, so that the experiencer when he opens his eyes sees the outer world transfigured."*

While Posford pointed out that he'd only used DMT three times in his life:

> *"...if listening to the music could achieve the same affect as a chemical hallucinogen it would be a powerful thing and something worth striving for."*   *(Grey, 2005)*

It seems that a great many artists have continued to strive for this. Finally, his statements also help explain the appeal of Shpongle. It's a music with which travellers – of hyperspace and physical space, psychonauts and event-goers – can identify.

> *"I would say that quite a large percentage of our audience appears to have certainly had that experience, and I think that it provides a way to relate. Our music creates a common thread and instant bond of alliance to other people who have had a psychedelic experience, in the same way that, say, traveling might. I think that I get on better with people if they've done psychedelics and traveled, because it opens your mind up in a way that is unequivocal. It makes one adept at relating and interacting in a playful, intangible, broadminded way that perhaps you don't have with people that maybe haven't had those experiences."*   *(Posford in Brown, 2011)*

The beneficence extended a long way. Space Tribe brothers Olli and Miki Wisdom were privy to portions shared from the magic bag in Anjuna, 1995. According to Olli: "DMT opened my eyes to a whole new universe... after that everything is perceived differently". For Olli's brother Miki – chief designer of Space Tribe clothing and CD artworks, it was a "blast of inspiration". Miki seemed to speak for many, stating that:

> *"It was one of the highest, most wonderful points of my life. I never felt more alive than after that first ego-peeling, moment stretching, visually exhilarating step through the whirling mandala gateway to the land of the crinkle-cut synapse, & life has never been the same".*

In an early production, Space Tribe used the sonorous chirping of insects to replicate the experience within DMT-space, on 12-inch *Ultrasonic Heartbeat*, which features 'Cicadas On DMT' (1996).

Space Tribe based themselves in Australia where DMT would have a significant impact, especially following McKenna's tour in 1997. In his talks, McKenna shared the wisdom, already known among small circles of ethnobotanical experimentalists, that DMT could be harvested from alkaloids in local *Acacia*, which fuelled the inspiration of local bent on the idea that they could derive DMT from the *wattle*, the national floral emblem (and local designation for *Acacia*). On 'Burning Point' (*Sun Control Species—Unreleased*, 2004). Australian artist Andrew Davidson (Sun Control Species) dropped a McKenna sample pungent with the acrid vapour: "The national symbol of Australia is the wattle. It's an *Acacia*. The *Acacia* ecology of Australia is jammed with DMT." Layered with efforts to reproduce the sound effects of DMT, the act Insectoid (Mark Turner, Nik Spacetree and Ray Castle) was a vehicle of influence, especially their album *Groovology of the Metaverse* (1998). 'New Vistas' from that album offers the voice of Alex Grey:

> *"I feel that I am merely an agent, giving you some keys, which have been given to me to pass on to you. These keys are to unlock doors out of your present prison. Doors opening in on new vistas. Doors beyond where you are now."*

Otherworldly tracks using exotic birdlife, insects (especially cicadas) and Aboriginal song lines – e.g. on 'Insecticide' and 'Tribedelic Nomads' – evoke the break-through into a DMT vista.

Throughout the last decade, a DMT-blend first prepared and popularised in Australia called *changa* has grown popular – and is now smoked on dancefloors around the world. *Changa* is a smoking mixture containing ayahuasca (*B. Caapi*) vine and/or leaf, and a combination of several admixture herbs and naturally sourced DMT (such as *Acacia*) infused into these herbs. This short-lasting preparation – using herbs like mullein, peppermint, pashion flower and blue lotus – has inspired other DMT-enhanced leaf blends including MAO inhibitors rendering the experience a "smokeable *ayahuasca*" (Palmer 2011/12).

The soundscapes of psychedelic electronica amplify DMT's diverse influence. At one end of the spectrum, DMT is enlisted in a kind of sonic arms war, where warrior DJs will display a psychedelic braggadocio and cavalier sampledelics care of, for example, Mister Black, whose 'DMT Molecule' breaks down with a monologue from American comedian Joe Rogan: "you should all smoke DMT and join my cult mother fuckers!" While this kind of material is common, a growing reverence for DMT is recognisable in work produced since McKenna's passing. This material evokes a yearning for the mystical experience, the desire to trespass across barriers that separate life and death, consciousness and unconsciousness, achieved, for better or worse, without a life of meditation training. Within the visionary arts and music community, especially among those who claim to have "broken through", DMT has been associated with a movement towards a state of grace, in which one has reconciled to the truth that what we know as death and life are not binary opposites. Shpongle had the measure of this state of grace on *Nothing Lasts... But Nothing is Lost* (2005) on which McKenna had the final word. The track "...But Nothing is Lost" is both a dirge sung for McKenna and an acceptance of impermanence offering his master's voice:

"Nothing lasts... nothing lasts. Everything is changing into

something else. Nothing's wrong. Nothing is wrong. Everything is on track. William Blake said nothing is lost and I believe that we all move on."

"Life must be the preparation for the transition to another dimension", explains McKenna on 'Molecular Superstructure' from the same album. As is conveyed by this remixtical artifice, with the expansion of personhood enabled by DMT, the barrier that separates life from death grows ambiguous.

Much of this material is expanded upon in my book forthcoming with North Atlantic Books, *Mystery School in Hyperspace: A Cultural History of DMT.* One of the key themes addressed in that book is the meta-ritualised character of the DMT experience; that it is a superliminal *initiation* device. Material in this article could serve as evidence for what Joseph Campbell called the "hero's journey". The DMT experience can be conceptualised as a hero's journey, where the user returns from the journey into hyperspace with gnostic revelations that inform culture in the form of stories (trip reports), music, art. Psychedelic electronica and its associated visionary arts community has received these visitations and is infused with a variety of storied encounters, adversities, ontological challenges in alien landscapes. The experience may be traumatic or inspirational, but more often than not, users overcome challenges to their own ontological security and indeed their own sense of mortality, in boon-like revelations that, as in McKenna's own telling examples, are brought back for the benefit of themselves and those around them. As its vapour suffuses soniculture, and its story is mixed into psychedelic fiction, DMT is harnessed as a gnostic technology, an instrument of revelation, facilitating a potent, if uncertain, initiation.

Brown, D. J. (2011). *Shpongle & Psychedelics: An Interview with Simon Posford.* http://mavericksofthemind.com/simon-posford.

Grey, A. (2005). Simon Posford: Synesthesia. *CoSM: Journal of Visionary Culture.* Winter Solstice. 16–18.

Hancock, G. (2006). *Supernatural: Meetings With the Ancient Teachers of Mankind.* Disinformation Books.

Luke, D. (2011). Discarnate Entities and Dimethyltryptamine (DMT): Psychopharmacology,

Phenomenology and Ontology. *Journal of the Society for Psychical Research*, 75(902), 26–42.

McKenna, D. (2012). *The Brotherhood of the Screaming Abyss: My Life with Terence McKenna*. North Star Press of St. Cloud, Inc.

McKenna, T. (1993). *True Hallucinations: Being an Account of the Author's Extraordinary Adventures in the Devil's Paradise*. Harper, San Francisco.

Palmer, J. (2012). Changa. *Entheogenesis Australis Journal* 3: 98–101.

Schultz, M. (2010). *DMT: The Spirit Molecule*. Spectral Alchemy.

Stevens, J. (1987). *Storming Heaven: LSD and the American Dream*. Paladin, London.

St John, G. (2012a). *Global Tribe: Technology, Spirituality and Psytrance*. CT: Equinox, Sheffield & Bristol.

St John, G. (2012b). Freak Media: Vibe Tribes, Sampledelic Outlaws and Israeli Psytrance. *Continuum: Journal of Media and Cultural Studies*. 26 (3): 437–447.

St John, G. (2013a). Writing the Vibe: Arts of Representation in Electronic Dance Music Culture. *Dancecult: Journal of Electronic Dance Music Culture* 5(1): http://dj.dancecult.net/index.php/journal/article/view/164/188.

St John, G. (2013b). Aliens Are Us: Cosmic Liminality, Remixticism and *Alien*ation in Psytrance. *Journal of Religion and Popular Culture*. 25 (2): 186–204.

St John, G. (2014). Sampling Religion: Electronic Dance Music Events, Media Shamanism and Superliminality. Lyden, J. & Mazur, E. M. (eds.) *The Routledge Companion to Religion and Popular Culture*. Routledge, New York. (forthcoming Oct).

Strassman, R. (2001). *DMT: The Spirit Molecule. A Doctor's Revolutionary Research into the Biology of Near-Death and Mystical Experiences*. 3rd ed. Park Street Press, Rochester, VT.

Discography

1200 Mics. (2002) *1200 Micrograms*. CD, Album. TIP.World: TIPWCD21.

Cosma Shiva. (2001) *In Memory Of Terence McKenna*. Vinyl, 12-inch. Shiva Space Technology: SST EP 007.

Dark Nebula. (2003) *The 8th Sphere*. CD, Album. Digital Psionics: DPSICD06.

Eat Static. (1993) *Abduction*. CD, Album. Planet Dog: BARK CD 001.

Emok & Banel – *Nuance*. (2006) CD, Compilation. Iboga Records: IBOGACD43.

F.F.T. (2006) *Future Frequency Technology*. CD, Album. AP Records: AP152.

Insectoid. (1998) *Groovology of the Metaverse*. CD, Album. WMS Records: WMSLP02.

Jimi Hendrix Experience, The. (1967) *Are You Experienced*. LP, Mono. Track Records: RS 6261.

Patchbay. (2012) *Southern Cross*. CD, Album. Mosaico Records: MSCRCD001.

Ritmo. (2011) *Phrase A*. CD, Maxi-Single. Iono Music: INM1MCD003.

Sphera – *Wide Open*. (2011) 9 × File, WAV, Compilation. Echoes Records: ECHOEP-IL056.

Shpongle. (1998) *Are You Shpongled?* CD, Album. Twisted Records: TWSCD4.

Shpongle. (2000) *Divine Moments of Truth*. CD, Single. Twisted Records: TWSC14.

Shpongle. (2005) *Nothing Lasts... But Nothing is Lost*. CD. Twisted Records: TWSCD28.

Space Tribe. (1996) *Ultrasonic Heartbeat*. Vinyl, 12-inch. Spirit Zone Recordings: SPIRITZONE016.

Streamers. (2012) *Power and Light*. 3 × File, MP3, EP. Yellow Sunshine Explosion: YSEDEP001.

Sun Control Species. (2004) *Sun Control Species – Unreleased*.

PSYCHEDELIC MINDFULNESS: CAN MEDITATION BE USED IN CONJUNCTION WITH PSYCHEDELICS TO INCREASE SELF-AWARENESS?

ALEXANDER BEINER

Why do people take psychedelics or meditate? There are probably as many answers to this question as there are people. Personal growth, self-medication, religion, curiosity and recreation are perhaps the most common reasons. As interesting as the others are, it is personal growth that I will be looking at in this article. Personal growth is a nebulous term, however. It is usually understood as a general increase in the qualities and abilities that we generally view as psychologically desirable, including emotional intelligence and cognitive function.

As a result, I will be narrowing the parameters to focus purely on self-awareness, as it is an essential quality in all psychological and spiritual growth. Specifically, I'm interested in whether the use of psychedelic substances and the practice of mindfulness meditation can be used in conjunction to reliably increase our self-awareness, and how this might look in practice.

In order to know whether an individual is becoming more self-aware through either or both of these practices, we would first need to decide how to measure self-awareness in the context of personal development. There are a number of models available that measure personal development in general, including Spiral Dynamics and the Johari Window. Spiral Dynamics argues that human development evolves through an infinite number of stages, moving from a basic stage where an individual is driven only by instinct, to a 'Global View' comprising a more inclusive and holistic outlook, with individuals moving through stages and gaining awareness by increasingly accepting, or integrating, previous stages to create more inclusive, empathetic perspectives on themselves and the world. (Beck, 2003). Similarly, the Johari Window measures how much a person is aware of in themselves and others and helps individuals to work on themselves or collaborate together (Luft, 1955).

Both approaches revolve around self-awareness to a large degree, and both have their uses. However, for an understanding of how different practices and therapies affect consciousness on a phenomenological level, I find Ken Wilber's work around Integral Theory to be of most use, not least because his later work draws heavily on Spiral Dynamics. However, it is his early work on the 'Spectrum of Consciousness' that is of most use to us in this instance.

FIGURE 1 (Wilber, 2001) demonstrates the model in question. It represents different levels on the spectrum of self-awareness, determined by where we 'draw the line' between self and other. Mapped against each level (Personal, Ego, Total Organism, Unity Consciousness) are various therapies that apply to recognising and letting go of the diagonal line cutting one part of our awareness off from another.

For example, an individual's sense of self might be limited purely to a narrowly defined persona, split off from the rest of the ego and from their unconscious urges. Someone struggling with a split on this

Therapies and the levels of the Spectrum

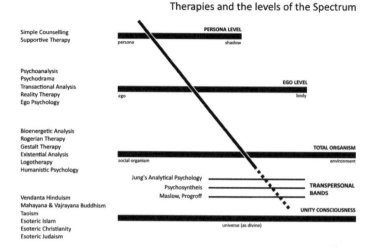

'Persona Level' may draw the line of 'me and not me' through their own psyche by deciding that all altruistic thoughts in their own head are good and acceptable (me), while all selfish and angry thoughts are bad and unacceptable (not me). In this way only a portion of the psyche is accepted and recognised as 'me', whereas all natural thoughts and impulses of selfishness or anger are suppressed and then projected as 'not me'. (ibid.)

Conversely, another person may be aware of and able to integrate all these differing aspects of their own psyche, and also have healed the rift between mind and body, seeing them as interrelated rather than separate. As well as mind and body becoming whole, an individual might explore a therapy which helps to remove the distinction between their entire organism and their environment.

Note that at the bottom of the graph the dividing line begins to break apart and then disappear. These are the transpersonal states between what Wilber refers to as the 'Total Organism' and unity consciousness, in which we perceive ourselves to be one with all of reality. Mystical

FIGURE 1: Functional segregation in the visual cortices

traditions around the world tend to concern themselves with leading people to this state of awareness.

However, it is vital to understand that 'lower' levels do not mean 'better' levels. As shown above, different therapeutic and spiritual practices are designed to dissolve boundaries at different levels, and each individual may set the boundary in each level differently at different times. (ibid.) That is why it is not unheard of for a long-term meditator, or someone who is expert at navigating psychedelic experience, to still have an unresolved split between the ego and the persona or the mind and body despite having experienced unity consciousness. For example, someone might meditate for 50 years but still be homophobic. Similarly, a person who uses psychedelics for personal growth might nevertheless cling to an illusory boundary between 'us' in the psychedelic community and 'them' outside of it, and even be strengthening these internal boundaries through their psychedelic use. In short, having a mystical experience of total unity with the rest of reality does not make you a wise or grounded person.

And so we return to the original question; if there are various methods for dissolving these illusory boundaries at different levels, and everyone draws the line between self and other differently, how does this affect how we can use psychedelics in conjunction with mindfulness meditation? Is it possible to work on all of these levels using these tools, rather than just the transpersonal levels we tend to associate them with?

To see how these two practices might be used together across the whole spectrum of our awareness, it's necessary to delve into what effect they have on self-awareness in isolation, beginning with mindfulness meditation. This form of meditation can be defined as objectively observing your own sensory experience without reacting to it. It is a very simple practice that trains the individual to first develop a distinction between the watching subject and the 'objects' within their experience. This gives one the ability to observe one's thoughts, emotions, sensations and body without resisting, grasping, judging or running away. It has been shown to be effective in treating a number of psychological illnesses, including reducing anxiety and depression (Kabat-Zinn et al, 1992).

It can also improve concentration (Lutz et al, 2009) empathy (Mascaro et al., 2012), emotional awareness and regulation (Lutz et al., 2008).

As impressive as the effects of mindfulness meditation may be, a growing body of evidence now exists to show that psychedelics can increase our self-awareness and indeed heal many psychological illnesses. This includes psilocybin for the treatment of OCD (Moreno et al., 2006), Ayahuasca for treating anxiety (Santos et al., 2007), and Ketamine for the treatment of alcoholism (Krupitsky and Grinenko, 1997). Added to this, thousands of years of shamanic practices leave no doubt that psychedelics reliably elicit experiences that have a profound effect on an individual's self-awareness. A participant in a recent study into the effect of LSD psychotherapy for anxiety associated with life-threatening illness (Gasser et al., 2014) recounted:

"My LSD experience brought back some lost emotions and ability to trust, lots of psychological insights, and a timeless moment when the universe didn't seem like a trap, but like a revelation of utter beauty."

(Sullum, 2014)

Interestingly, both mindfulness meditation and some psychedelics have a similar effect on the brain. fMRI scans of the brains of long-term meditators have demonstrated that meditation leads to a decrease in activity in a number of brain regions, notably the Default Mode Network (DMN). Brewer et al. (2011) write:

"We found that the main nodes of the default-mode network (medial prefrontal and posterior cingulate cortices) were relatively deactivated in experienced meditators across all meditation types."

A recent fMRI study of subjects who had taken psilocybin demonstrated strikingly similar observations, namely a decrease in activity in the DMN (Carhart-Harris et al., 2012). Carhart-Harris et al discuss the relevance of this decreased activity:

"The high metabolic activity of the PCC and the default-mode network (DMN) with which it is associated has led some to speculate

about its functional importance, positing a role in consciousness and high-level constructs, such as the self or 'ego'. Indeed, the DMN is known to be activated during self-referencing and other high-level functions linked to the self-construct."

By extension, this decrease in activity in the DMN may explain some of the subjective experiences of meditators and people who use psychedelics. While the question remains of how exactly these physical changes influence our subjective experience of both states, it is possible that this correlates to the experience in both meditators and those experiencing a psychedelic state of a dissolution of boundaries, and an 'opening' of what Aldous Huxley referred to as the 'reducing valve' of the conscious mind that blocks out the majority of our sensory and emotional experience (Huxley, 1954). When the 'valve' is opened, the normal mode of consciousness in which the internal narrator is constantly creating and becoming attached to illusory boundaries is temporarily suspended. The result is an experiential awareness of the arbitrary nature of these boundaries and a sense of unity, wholeness and even bliss. Or as the Fourth Century Zen monk Sengzhao put it: "Heaven and earth and I are of the same root, the ten thousand things and I are of one substance" (Suzuki, 1959).

Neurologically, this experience of no boundaries corresponds to the illusory nature of the narrating subject in the mind and brain. As Dr Rick Hanson points out:

"...a unified, enduring, independent I who is the essential owner of experiences and agent of actions just doesn't exist... The brain strings together heterogenous moments of self-ing and subjectivity into an illusion of homogenous coherence and continuity. The self is truly a fictional character." *(Hanson and Mendius, 2006)*

Meditation and psychedelics both have this effect of allowing this illusory self to dissolve and increasing our awareness of our own self-constructed boundaries. So, why not just use one or the other for personal development? The answer, I believe, can be found in the differences in phenomenology between the two experiences and the huge difference between the meditative state and the psychedelic state.

Meditation is a process of slowly and methodically paying attention to conceptual boundaries until they are, simply through the act of observation, revealed as illusory. In Zen, students are often instructed to ask the question 'who is sitting?' or 'who am I?' while they meditate. This is not to lead them to an answer, but rather to allow them to see that there is no answer. By asking the question the student sees that it is impossible for the subject to be the subject of the subject; the person asking the question is the answer. Note that there is no attempt to destroy the boundary, because this would imply it was there in the first place. Instead, meditation encourages the student to simply observe these boundaries until noticing that, like Dr Hanson's 'illusory self', they never existed in the first place.

In this way, mindfulness meditation shows the practitioner that striving to create and maintain boundaries is both futile and ultimately impossible, leading to an experience of calm and peace. There is quite literally nothing to worry about, because we create our own neurosis. This isn't to say that these boundaries aren't *experienced* as real or that we don't suffer, or won't continue to suffer because of them, simply that they are arbitrary and can be changed.

A useful metaphor is to think of the border between the USA and Mexico. This border is seen as real by millions, to the extent that thousands have died to defend it or overcome it. However, it is also 'not real' in the sense that it is an arbitrary line that relies entirely on thoughts and ideas to survive. Birds and coyotes do not line up at the border to show their passports, because they do not recognise imaginary barriers, only physical ones.

The psychedelic state dissolves barriers in a very different way. Instead of requiring self-motivation and discipline to slowly and methodically observe and detach, psychedelics *force* boundary dissolution in a very short space of time. This kind of explosive awareness occurs very rarely, if at all, during a meditation practice, and it's an important distinction. Letting go of our boundaries is a terrifying thing to do, and even if we're trained to meditate on them we often maintain them rather than risking the unknown.

However, once you take a high enough dose of a psychedelic, this boundary dissolution will happen whether you want it to or not. This difference in the freedom of choice really is crucial, because there are occasions, such as addiction and severe depression, where the individual needs an extra push to start dissolving old patterns. Given free choice, we often choose to not deal with something potentially painful, even while we're meditating. I refer to this maintaining of boundaries using meditation as 'taking a long hard look in the other direction' as students, myself included, sometimes use the ability to concentrate to avoid certain issues with more intensity than we could have without knowing how to meditate.

Overcoming this through forced dissolution is probably the most useful aspect of psychedelics in expanding our self-awareness. There is, however, another factor. The psychedelic state can also illicit another crucial aid to psycho-spiritual growth that can be more difficult to experience through meditation, namely an encounter with the 'Other', or a connection to a third-person, transpersonal intelligence. I am not making any ontological statements about this third-person intelligence, simply pointing out that phenomenologically it is a common aspect of both experiences, but far more easily accessed using psychedelics, regardless of the stage of development of an individual.

This is important because it is often necessary to have an outside perspective on oneself in order to develop a healthy level of awareness without adding unnecessary boundaries. This may be a therapist or instructor, or in Eastern meditation traditions a guru. However, because these are inevitably other people with their own psychological baggage and agendas, they also run the risk of giving us more psychological baggage to deal with. In my opinion it is safer and more reliable to use the third-person 'Other' encountered in the psychedelic experience as a guide for personal growth as it is intimately linked to the individual 'communicating' with it.

And yet, despite these benefits, the psychedelic state at the same time has its limits in expanding our self-awareness. Set, setting and dose play a defining role in how well someone can navigate their experience and can interfere with someone's ability to explore their consciousness. Even more importantly, what we get out of a psychedelic experience is limited by our own ideas and cultural baggage. Wilber points out that, as well as stages of psycho-spiritual development, there is a range of states we can all experience regardless of who, where or when we are. These include waking, dreaming, the psychedelic state or the meditative state, hypnogogic and everything in between. States are open to all people at all times; anybody, regardless of what stage of development they are currently in, can experience a psychedelic state, even a peak mystical experience. However, "they will always interpret this state based on the contextual framework of their stage of development" (Wilber, 2006). You can experience any state you want, but none of us experience it in the same way throughout our lives, and you will always be limited in how much you can understand from the state based on your stage – your cognitive intelligence, worldview and overall awareness (Wilber, 2006).

What is needed, in that case, is a practice that moves individuals up stages faster, so that they can interpret their states more effectively and use these insights to move up the stages of the Spiral Dynamics model. Not surprisingly, meditation can do exactly this. Wilber writes:

> *"Considerable research has demonstrated that the more you experience meditative or contemplative states of consciousness, the faster you develop through the stages of consciousness. No other single practice or technique – not therapy, not breathwork, not transformative workshops, not role-taking, not hatha yoga – has been empirically demonstrated to do this. Meditation alone has done so. For example, whereas around 2% of the adult population is at second tier [in Spiral Dynamics], after four years of meditation, that 2% goes to 38% in the meditation group."* (Wilber, 2006)

We have also seen that mindfulness meditation and the use of psychedelics have benefits and drawbacks when used in isolation to increase self-awareness. Indeed, the very thing that distinguishes them also limits

their effectiveness in isolation. Psychedelics force boundary dissolution that might not have happened, and meditation requires willpower to achieve a similar state without boundaries, but can also be used to reinforce boundaries. Likewise, psychedelics can force these unhelpful boundaries to dissolve – but only if the individual is at a stage where they can usefully contextualise the experience. We have no evidence yet to suggest that psychedelics will move you up these stages reliably, whereas meditation can.

Added to this, meditation improves our overall well-being and increases concentration, as well as training us to gradually and safely practice boundary dissolution. This makes it the ideal practice to compliment the psychedelic state. The heightened awareness and acceptance of impermanence gained through meditation, and the mental discipline and concentration, are incredibly useful during a psychedelic experience. Being able to stay objective and centred as your identity, environment and perceptions become fluid is essential in navigating the psychedelic state, and not 'freaking out' as some psychonauts would put it.

As a result, I believe an integrated approach of mindfulness meditation combined with the responsible use of psychedelics is an

FIGURE 2: TTI (Train -Trip -Integrate)

excellent way to reliably and safely expand self-awareness. So how might this 'Psychedelic Mindfulness' look in practice? I propose a three-part process called TTI (Train – Trip – Integrate) which I have outlined in FIGURE 2.

The first stage is to learn to meditate, and in doing so become more mindful of yourself and your environment. Recent studies suggest it takes around 11 hours of mindfulness meditation practice to begin eliciting physical changes in the brain (Yi-Yuan Tang et al., 2010). In my experience as a meditation trainer this practice is best attained intensively using an integral, non-religious framework. These hours can also be attained with 20 minutes of daily practice for around five weeks, but having trained people both through six-week courses and over an intensive weekend, I have observed that the intensive method is considerably more effective. After these 11 hours have been attained, a daily practice or around 20 minutes a day to start with maintains the skill. With practice, the skill changes from a state to a trait – from a momentary experience to a state of being.

The next stage is the 'Trip' stage. Ideally, before going into the trip one would have engaged in some multi-disciplinary study of the effects of psychedelics, including set, setting and dose, relevant shamanic and modern healing modalities. With a healthy context, a psychedelic could then be taken in either a ceremonial or natural setting with a sitter or facilitator, with the intention of increasing self-awareness as opposed to recreation.

After the experience one would continue to practice meditation for 15 or 20 minutes a day to objectively reflect on the insights gained and implement the necessary changes in thought and behaviour highlighted by the psychedelic experience. Mindfulness meditation is the ideal tool for this integrative process because of the importance of acceptance, non-judgement and self-compassion in the practice. I believe there is still much to do to create a supportive context in which individuals could practice this kind of integration using meditation, but that it is both possible and desirable.

Once the insights have been integrated and our boundary lines 'extended', the process could be repeated as demonstrated in FIGURE 2, leading to continued personal growth. This is very much a provisional model and requires further research, as well as a more structured programme for the individual to follow that draws on the latest insights in psychedelic psychotherapy, but I strongly believe that this type of 'Psychedelic Mindfulness' could eventually become one of the most powerful tools we have for personal development, healing and happiness.

Beck, D. (2003). *Spiral Dynamics, Mastering Values, Leadership, and Change.* Blackwell Publishing, Malden, MA.

Brewer, J. et al. (2011). Meditation experience is associated with differences in default mode network activity and connectivity. *PNAS* 2011 ; published ahead of print November 23, 2011.

Carhart-Harris, R. et al. (2012). Neural correlates of the psychedelic state as determined by fMRI studies with psilocybin. *PNAS* 2012. Published ahead of print January 23, 2012.

Gasser, P. et al. (2014). Safety and Efficacy of Lysergic Acid Diethylamide-Assisted Psychotherapy for Anxiety Associated With Life-threatening Diseases. *The Journal of Nervous and Mental Disease*, March, 2014.

Hanson, R. & Mendius, R. (2009). *Buddha's Brain: The Practical Neuroscience of Happiness, Love & Wisdom.* New Harbinger Publications, California. 2009. 214.

Huxley, A. (1954). *The Doors of Perception.* Chatto and Windus.

Kabat-Zinn, J. et al. (1992) Effectiveness of a meditation-based stress reduction program in the treatment of anxiety disorders. *Am J Psychiatry*, 1992. 149 (7), 936-43.

Krupitsky E. M. & Grinenko A. Y. (1997). Ketamine psychedelic therapy (KPT): a review of the results of ten years of research. *Journal of Psychoactive Drugs*, 1997. Vol. 29, N.2, pp.165-183.

Luft, J. & Ingham, H. (1955). The Johari window, a graphic model of interpersonal awareness. *Proceedings of the western training laboratory in group development.* UCLA, Los Angeles.

Lutz, A., Brefczynski-Lewis, J., Johnstone, T. & Davidson, R. J. (2008). Regulation of the Neural Circuitry of Emotion by Compassion Meditation: Effects of Meditative Expertise. *Plos One* 2008 March. 3(3). e1897.

Lutz, A. et al. (2009). Mental Training Enhances Attentional Stability: Neural and Behavioral Evidence. *The Journal of Neuroscience*, 21 October 2009. 29(42).

Mascaro, J. S., Rilling, J. K., Tenzin Negi, L., & Raison, C. L. (2012). Compassion meditation enhances empathic accuracy and related neural activity. *Social Cognitive and Affective Neuroscience.* 2012, 48-55.

Moreno, F. A., Wiegand, C. B., Taitano, E. K. & Delgado, P. L. (2006) Safety, tolerability and efficacy of psilocybin in 9 patients with obsessive-compulsive disorder. *J Clin Psychi,* 2006. 67: 1735-1740.

Santos, R. G. & Landeira-Fernandez, J., Strassman, R. J., Motta, V. & Cruz, A. P. M. (2007). Effects of Ayahuasca on psychometric measures of anxiety, panic-like and hopelessness in Santo Daime members. *Journal of Ethnopharmacology,* 2007. 112 (3): 507-513.

Sullum, J. (2014). First Study Of LSD's Psychotherapeutic Benefits In Four Decades Breaks Research Taboo. Available: http://www.forbes.com/sites/jacobsullum/2014/03/04/first-study-of-lsds-psychotherapeutic-benefits-in-four-decades-breaks-research-taboo/. Last accessed 11 March, 2014.

Wilber, K. (2001). No Boundary: Eastern and Western Approaches to Personal Growth. Shambala Publications, California. 2001. 14-15

Wilber, K. (2006). *Integral Spirituality.* Integral Books, London. 196-197.

Tang, Y., Lu, Q., Geng, X., Stein, E. A., Yang, Y. & Posner, M. I. (2010). Short-term meditation induces white matter changes in the anterior cingulate. *PNAS* 2010 ; published ahead of print August 16, 2010.

PSYCHEDELIC DHARMA

ALLAN BADINER

It has been over a decade since *Zig Zag Zen: Buddhism and Psychedelics* (Badiner, 2002), a book that I edited and for which Alex Grey edited the art, was published. Based largely on a special section on the subject in *Tricycle: The Buddhist Review*, in the fall of 1996, the book, published by Chronicle Books, was the first of its kind in terms of closely examining the relationship between psychedelics and Buddhism through a variety of voices and perspectives. Each of the 33 contributors had a different story to tell... different substances, different circumstances, different experiences, and different lessons drawn.

For some, it was a story about giving in to temptation, and suffering a lack of clarity, or worse, lasting confusion and addiction. For others, it was a brief glimpse into a rarefied world of intense sense impressions accompanied by fanciful but useless imagery. For a few others still, it was the very threshold of their journey into the truth of Buddha's teachings, with unforgettable, if fleeting, insights. All the contributors had a significant story to tell, and each person's facet of the truth of drugs and Dharma remains riveting and revealing.

My 25-year fascination with Buddhism is primarily about how it shows up in the challenges of modern life, how it informs our struggles with social problems. My earlier books focus on Buddhism in relation to economics and globalisation (*Badiner*, 2002), and Buddhism and the ecological crisis (*Badiner*, 1990).

But why psychedelics? Certainly, the use and abuse of drugs are a giant social and legal problem – created of course mainly by those who think it is a problem, or want it to be a problem. The structures and thinking in place to control drug use in our society may actually drive the lion's share of the abuse and inappropriate use.

While psychedelics lurk in the personal histories of most first-generation Buddhist teachers in Europe and America, today we find many teachers advising against pursuing a path they once travelled. The fifth precept, the Buddhist admonition against drinking intoxicating liquors, is frequently touted as the end of conversation about using psychedelics regardless of the sincerity or intensity of higher purpose motivating such journeys (Warner, 2010).

In fact, the intentional use of consciousness expanding psychedelics for personal growth purposes has little in common with - for example- imbibing a bottle of gin. Buddhism and psychedelics share a great deal in common: an acute attraction to mindfulness and present moment awareness, and the primacy of direct experience. In their purest applications, they share the same ultimate purpose: the attainment of liberation for the mind.

Few Buddhists make the claim that psychedelic use is a path itself, some maintain that it is a legitimate gateway, and others – acutely aware of the inconvenient duality that undermines any chemically dependent spiritual path – feel Buddhism and psychedelics don't mix at all. But just as Buddhism itself – according to its founder – must be held to the test of personal experience and to the wholesomeness or unwholesomeness of the results, so also must the question of how, or if, psychedelics can be a legitimate part of a Dharma practice. Buddhism is a path of freedom "visuddhimagga" (in Pali) and so we are all free – in fact, must be free – to make these choices for ourselves.

Encouraging critical examination and analysis, and the freedom to make these discoveries for oneself, is an essential foundation of Buddhism and is found as far back as the Kalama Sutta. The Kalamas lived in an area of NW India called Kesaputta, and were primarily of the merchant class. The Buddha was advising them on how to verify truth. I think the Buddha's advice to them bears quoting in some detail:

> "'Do not go upon what has been acquired by repeated hearing,' says the Buddha. '...nor upon tradition; nor upon rumor; nor upon what is in scripture; nor upon one's surmise; nor upon an axiom; nor upon logical reasoning; nor upon a bias towards a notion that has been pondered over; nor upon another's seemingly keen mind; nor upon the consideration that this important monk is our teacher,' The Buddha warned the Kalamas, 'Only when you yourselves know in your own experience – that these things are good; that these things are not blamable; that these things are praised by the wise; undertaken and observed to lead to benefit and happiness in you – should you abide in them.'"
>
> (Bodhi, 1988)

Just as social prohibitions on what ideas to let in to your consciousness are anathema to Buddhism, so also must be such restrictions on what plants to let in. Tibetan Buddhists, for one, have developed over the centuries a wide field of psychopharmacology and have an endless number of psychiatric botanical medicines – practically none of which have been previously identified or scientifically tested in the West (Clifford, 1994). Laws against psychedelics, many of them potentially helpful, create a catch-22: they are banned because they "have no accepted medical use," but researchers cannot explore their therapeutic potential because they are banned.

When I use the term psychedelics – also called entheogens, an accepted word despite my essentially agnostic Buddhist perspective – I am referring to the use of plant materials to expand consciousness and trigger primary spiritual experience. The problems caused by cocaine, heroin, GHB, ketamine, methamphetamines, and other consciousness-constricting drugs are indisputable and, as far as I'm concerned, indefensible. The notion that all 'drugs' are fundamentally alike is at the root of the confusion in our drug laws and the social debate about them. Drugs differ. Uses and occasions differ. Policies and practices also ought to differ appropriately.

Drug use will always be with us, so therefore it seems reasonable that recreational users of psychedelics who are also responsible users should be treated more or less the same way recreational drinkers are. Medical users of psychedelics should be treated the same as anyone using commercial psychopharmaceuticals to protect their health. Spiritual users of psychedelics should be respected as any seekers of greater wisdom desiring shifts in consciousness. Too slowly are we collectively grasping the reality that abuse of dangerous drugs is less of a legal issue than a medical one.

The use of psychedelics as sacrament is well known throughout history (Merkur, 2001), but it is interesting how it intersects with Buddhism now at a time when both practices are prominent at the base of a massive and burgeoning human quest for greater understanding of the self. Until just the recent past, awareness about the deepest 'occult' or 'hidden' parts of our spirit selves was considered the private preserve of shamans, priests, or spiritual masters who had 'earned' access to it. Religious experience was mediated by these authorised few, and this is a tradition still with us in the form, if not attitude, of many religions to this day. The democratisation of psychedelics, however, and of Buddhism to a similar extent, has been very much about the breakdown of this restricted access to the divine. In Buddhism, as in psychedelics, the individual takes responsibility for their relationship to the source of their being, and for access to the highest states of spirit mind.

Awareness of the relatedness between people, and between objects, and particularly the intimate understanding of the interrelated nature of opposites, are among the key insights psychedelic travellers often bring home from their plant-assisted 'pilgrimages.' Perhaps the popularisation of both Zen and psychedelics has had a major role in shifting the cultural mind from a dominantly Cartesian linear view of reality toward a mode of awareness that is more ecological and holistic. While we will always continue to think in linear ways, awareness is growing that it is relative, a human construct, and not a reflection of some 'objective reality'.

This way of seeing is not something people necessarily need psychedelics to experience. It is, in fact, one of the central premises underlying Zen and its doctrine of non-duality (Bodhi, 1994). This emergent worldview brings us closer to a perspective that is perhaps equally comfortable being called 'dharmic' or 'psychedelic'.

Putting aside myriad well-founded arguments both for and against psychedelic use, there is an essential Buddhist response to the long entrenched, ongoing, and devastating so-called war on drugs: great compassion (Edwards, 1998). I'd like to highlight this issue a little more... Despite the apparent falling of the Berlin Wall of Prohibition, draconian drug laws still ensnare millions of otherwise law-abiding people in an ever growing spiral of wasteful and counterproductive strategies, the foundation of which is punishment.

It has resulted in an incarceration rate so unimaginable that one in five of all people behind bars in the entire world are locked up in the United States (Mauer, 1997). At this very moment, American jails and prisons hold tens of thousands of people – vastly disproportionate numbers of them people of colour – whose only crime is possession of a plant. Prisons become classrooms for more advanced crime, drugs are readily available to everyone from school children on up, criminals outspend and outsmart police, and no one feels safer (Mason et al., 1997).

The drug war has led to cynicism and apathy and, of course, continues to blight thousands of lives. Profits from the illegal drug trade fuel organised crime and enhance the power of the cartels to corrupt police, judges, and government officials. The latest casualty in the failed war on drugs is our personal liberties. A society that actively banishes personal exploration with all psychedelic plants will need to closely monitor its citizens.

All our communications, transactions, and expressions are under dramatically increasing surveillance by a growing and expensive bureaucracy of control and repression. None of this is conducive to the peaceful and free contemplation of strategies for our personal liberation and fulfilment. In reality, this ceases to be a war on drugs, but rather becomes a war on consciousness, war on free exercise of that most precious of gifts bestowed on a human being.

Human history can arguably be seen as a series of relationships with plants, relationships made and broken. Plants, drugs, politics, and religions have harshly intermingled – from the influence of sugar on mercantilism to the influence of coffee on the modern office worker; from the British forcing opium on the Taoist Chinese to credible reports that the CIA used heroin in the ghetto to choke off dissent and dissatisfaction. The lessons to be learned can be raised into consciousness; integrated into social policy, and used to create a more caring, meaningful world, or they can be denied with the results now plainly seen.

The enhanced capacity for extraordinary cognitive experience made possible by the use of plant psychedelics may be as basic a part of our humanity as is our spirituality or our sexuality. The question is how quickly we can evolve in our collective maturity sufficiently so as to have informed and balanced examination of these issues.

As the eminent thinker Huston Smith is fond of pointing out: while psychedelic use is all about altered states, Buddhism is all about altered traits, and one does not necessarily lead to the other (Badiner, 2002). One Theravadin monk likened the mind on psychedelics to an image of a tree whose branches are overladen with low-hanging, very ripened, and heavy fruit. The danger is that the heavy fruit – too full and rich to be digested by the tree all at once – will weigh down the branches and cause them to snap.

On the other hand, Alan Watts, one of the first prominent Westerners to follow the Buddhist path, considered both Buddhism and psychedelics to be part of an ideal individual philosophical quest. He was not interested in Buddhism to be studied and defined in such a way that one must avoid 'mixing it up' with other interests and disciplines, such as in quantum theory, Gestalt psychology, aesthetics, or most certainly, psychedelics (Watts, 2013).

One time, when Alan Watts was giving an Esalen workshop, he was asked which way is the best path to enlightenment, meditation or psychedelic drugs. He chucked thoughtfully, and then said, "Well, I don't know about a 'best' way, but perhaps you could think of it like this, you can take the plane to New York, or you can walk."

While it is clear Buddhism and psychedelics inform each other, their intersection also makes room for much confusion. We ought to continue to have the conversation. Surely, it is only when we have the courage to see the truth in every other voice that we recognise the deepest and truest reflection of what is relatively real in our own.

Badiner, A. (1990). *Dharma Gala: A Harvest of Buddhism and Ecology.* Parallax Press, Berkeley.

Badiner, A. (2002a). *Zig Zag Zen: Buddhism and Psychedelics.* Chronicle, San Francisco.

Badiner, A. (2002b). *Mindfulness in the Marketplace: Compassionate Responses to Consumerism.* Parallax Press, Berkeley.

Bodhi, B. (1988). A Look at the Kalama Sutta. *BPS Newsletter.* #9, Spring '88.

Bodhi, B. (1994). Dharma and Non-Duality. *Buddhist Publication Society Newsletter.*

Clifford, T. (1994). Tibetan Buddhist Medicine and Psychiatry: *The Diamond Healing.* Motilal Banarsidass Publ.

Edwards, D. (1998). *The Compassionate Revolution: Radical Politics and Buddhism.* Green Books.

Mason, D., Birmingham, L. & Grubin, D. (1997). Substance use in remand prisoners: a consecutive case study. *BMJ* .315, #7099: 18-21.

Mauer, M. (1997). *Americans behind bars: US and international use of incarceration.* 1995. Washington, DC: Sentencing Project.

Merkur, D. (2001). *The Psychedelic Sacrament: Manna, Meditations, and Mystical Experience.* Inner Traditions/Bear & Co.

Warner, B. (2010). *Hardcore Zen: Punk Rock Monster Movies and the Truth about Reality.* ReadHowYouWant.com

Watts, A.. (2013). *The Joyous Cosmology: Adventures in the Chemistry of Consciousness.* New World Library

STONED TEMPLE PILOTS - SET, SETTING AND SUBSTANCE IN CONTEMPORARY ENTHEOGENIC SPIRITUALITY

JULIAN VAYNE

I want to speak today about a community of practitioners who are variously named and exist in Britain, throughout Europe and in many other lands. They are shamans, magicians, witches and followers of what we might call the 'medicine path'. These are a courageous group of people who are attempting to deploy an archaic method of spiritual enquiry that, in some cases, our culture currently legislates against in the strongest terms.

A recent case comes to mind:

> *"In September – November 2010 seven members of the Santo Daime church in the UK were arrested and placed on bail. Boxes*

of Santo Daime were seized and shown to contain DMT and a police investigation was launched. At this time charges were brought only against one person while the other 6 people remained on bail. The charge was Fraudulently Evading Prohibition on a Banned Substance. In these initial stages of the case the Defence legal team gathered evidence and expert witness statements to show the legal ambiguity of the status of ayahuasca, as well as the bona fide nature of the Santo Daime as a religious practice, and also the proven health benefits and lack of proof of harm concerning the sacrament. All this information was put before the Director of Public Prosecutions who then agreed to conduct a review of the case. The results of this review were that 5 people were released from bail without any further charges, but the DPP decided to pursue prosecutions against two individuals. In addition to the previous charges they were also charged with Conspiracy to import a banned substance."

There are many plants and chemicals being deployed by groups that are not currently prohibited in the nations within which they are being used. But where a violation of local law does occur the penalties can be dramatic. In the case of the two charged Santo Daime members they remained on bail for over two years. I am however pleased to say that, following magical work, spells and prayers that were deployed by much of the 'medicine community' (as I shall describe these 'practitioners') the State has chosen to drop the case for reasons of insufficient evidence. The fact that tens of litres of ayahuasca is deemed 'insufficient evidence' is clear proof of the power of magick.

I've been fortunate to encounter and, at times, share ceremonial space, with a variety of people within this 'medicine community' and while I've certainly never broken the law myself I have encountered ceremonial work that included the use of substances which are currently prohibited in the British Isles.

I have no idea how widespread these communities are. Considerations of security and probity are paramount and so I have no notion of the scale of these networks. However my sense is that there are perhaps thousands of medicine practitioners in the British Isles. This ranges from solitary practitioners who might identify as 'traditional witches' and who make use of herbs such as datura and mandrake, through to much more organised, communal religions such as the Santo Daime church.

So having set the scene and explained the limits of my knowledge I'd like to describe a few rituals. These are not by any means case studies but are best understood as phenotypic examples – composed of elements from different sources in order to create a composite of a ceremony's observable characteristics. They are entheogenic ritual mash-ups if you will.

The first ceremony takes place in a little village in Britain. The two participants call themselves witches, one man, one woman. The man has spent days brewing ayahuasca in his kitchen. He has sung and chanted in his own language as he made the brew; singing the famous icaro of Terrence McKenna, 'row, row, row your boat, gently down the stream...' while stirring the mixture. The plants, the vine and leaves, have been obtained from an online herbal supplier. He jokingly calls this 'boil in the bag' ayahuasca. The male witch is familiar with ayahuasca having taken it in his visits to South America. The female witch has never taken the potion before, she plans to use this as an opportunity to enter the spirit world through a new doorway (she has an intently animist view of the universe). The man wants to use this medicine to help heal the cancer of a family member.

Together they cast a circle using a simple Wiccan method. They call on the spirits of the four directions to protect them and inspire them. They both drink, and for the next two hours they sit. The woman is silent and motionless throughout, the man shakes a rattle, a constant beat – *tsk, tsk, tsk, tsk, tsk* – until the rite is done.

For the male witch when he finds himself in the right part of the vision he sees the cancer cells inside the body of the person he's seeking to heal. Strange spirit forms which look like aircraft, vaguely reminiscent of steampunk Wellington bombers, travel through the body of the patient. They train their guns on the diseased cells and blast them with rays of electricity.

At the end of the ceremony, the participants thank the spirit of the medicine from the land of the great river. Finally they go outside to breathe in the night air, to touch the bare earth and let their intentions go into the world.

As a supplement to this account, and to help you develop the charming metabelief that magick works, I'd like to add that the cancer was successfully cured.

Another ceremony. This time in a tepee. This one is clearly based on the peyote ritual of the Native American Church. There is a Grandfather peyote button on the crescent moon altar at the centre of the space, but the medicine is native psilocybin mushrooms (in honey). As with the previous ceremony the same rhythm – *dum, dum, dum, dum, dum* – but this time from a traditional water drum. The staff and rattle move round the circle, pulling the drum behind them, as each person who wants to gets a chance to sing.

Longtime members of the group know the Native American chants and sing out. Others know English Pagan chants and these are shared in the circle too. At times throughout the ceremony cedar is used to cleanse the space and tobacco is smoked as prayers are offered. The midnight water (a chance to stand up, leave the tepee and urinate as well as to have some water to drink) and the morning breakfast (deer meat, fruit and corn) serve to confirm that this is very much the peyote circle 'design' (a word these practitioners rather like) but deployed in a European context.

During the ritual the ceremonial fire is swept into astonishing patterns; a heart is created from the burning coals, or a great fiery bird. The fire is separated into two and brought back together again.

The leader of the ceremony explains the basic technique behind this complex ceremonial event to me:

> *"Especially when we are using peyote, 'the idea is quite different from psychoanalysis. Rather than dredging up our problems and trying to work on them we instead fill ourselves with joy. We fill ourselves with gratitude; we look to be conscious of the many, many ways in which*

we are blessed. In this way we push out the sadness, drive it away and open ourselves to the Great Spirit.'"

This kind of neo-tribal ritual is of course enacted in a much larger, though in many ways as deeply codified, ritual of the free party or rave.

In a city in the south west of England a warehouse has been over-run by outlaws. Inside the space, balanced on old trucks, is a vast sound system. The dark space is punctuated by florescent lights and UV glows. The deep drum pounds the rhythm, the heartbeat of the city – *duf, duf, duf, duf, duf.*

In a canopied chill-out area there are banners of Shiva Nataraja and images of the trumps from the tarot of Freda Harris and early entheogenic enthusiast Aleister Crowley. Here people are kissing, receiving massages, smoking and talking.

Cannabis and MDMA are the primary medicines in evidence, and while this may not be conceptualised by the majority as some form of spiritual practice the 'design' of the ceremony, its core technology, is clearly analogous to the peyote circle ceremony. Driving out the sadness with joy; with Ecstasy.

Allow me to tell another couple of stories.

The first is of a ceremony where ketamine is the sacrament. In this ritual a huge projection screen has been erected in a warehouse space, but this time the people at the party are by invitation only. On the screen a film is about to be shown, a computer simulation of the journey through the shushuma, the spinal column of Indian esoteric anatomy and the *axis mundi.* The explicit aim of the ceremony is to catalyse a direct experience of the non-dual nature of reality. Participants sit or lounge on the floor which is covered with fluffy blankets, cushions and the occasional cuddly toy (there are furry snakes and soft, friendly, white tigers).

Travelling through each chakra in turn, following the raising of the kundalini snake, the film contains beautifully crafted computer graphics. The human body is shown in space, vibrating with coloured light and morphing with patterns connected to this mythic journey.

The traditional Hindu symbols associated with each mystical centre are depicted; lotuses, Sanskrit letters and deities, as well as fantastic fractal graphics coloured with the hues associated with each area of the subtle anatomy.

Each participant will be offered a moderate dose of ketamine, via insufflation, before the film and again half way through the screening (as the serpent arrives at the heart centre). During the movie the person leading the ceremony sings the bija mantras at appropriate points and describes the attributes of each chakra, there is a background soundtrack of the mantra carrying participants through this experience which lasts approximately one hour. Those experienced psychonauts serving the medicine take only low doses and they are also on hand to help with any difficulties that may arise. This is a powerful act but conducted in a way that is as safe as possible.

The next ceremony is one in which participants gather outside in a circle of trees. There is a drum singing, I expect by now you know the rhythm...

There are voices too, calling on the Great Spirit, the life-force of the universe.

Participants spiral into the centre of the circle. As they do so they are imagining that they are walking across the vast swathes of time since the creation of our planet up until this present moment. Once assembled round the edge of the circle they go through a simple series of yoga type movements. Relaxing and opening their somatic selves for the experience yet to come.

At the centre of the space they partake of substances representing the four elements; honey for fire, scented herbs for air, water and the smell and touch of soil for earth. On the ground are blankets and animal skins. At the centre of the circle is where the participants who 'take the trance' sit (not all do so. Some remain drumming and singing on the perimeter of the circle, surrounding the blanketed area). Those taking the trance are each offered a large hit from a pipe of 5-MeO-DMT. This represents the element of spirit, the self-conscious coming into awareness of the life-force of our planet, the fifth element of this

Eucharistic style rite. Ritual chanting and drumming is kept going throughout the ceremony. As people who have smoked the 5-MeO-DMT emerge from the experience they too join in the chorus of voices.

At the end of the ritual, after everyone who has journeyed has come out of their trance, the participants thank the spirits of the place, they laugh and hug one another saying "good to see you Brother, good to see you Sister!".

So with these archetypal examples in mind what can we say about this 'medicine community'?

The first point is that, and this is my reason for including the rave in the examples above, they represent something active in the zeitgeist.

The notion that drug use might have for some (at a rave) or for all participants (as in the other examples) a spiritual dimension is far from alien these days in our culture. For while discourses about the harm and horror of drugs are commonplace so too is the idea that substances like LSD can fuel personal transformation and be allies in the creative processes of visual art, science, music, philosophy and more. We're also very aware through the media of both ancient and modern cultures in which use of sacred medicine is normal and accepted.

We live in a chemical world, and so the notion of spiritual experience as being something which might be augmented or even generated by chemicals is quite intelligible. This view of humanity is very new and quite different from much of the Western cultural tradition. It is perhaps easy to see it as an expression of a presumed archaic sensibility; namely that all things are connected, and that matter and spirit are not separate things. But whether this is a new or old idea really doesn't matter, the bottom line is that the possibility of psychoactives having a role in spirituality, religion makes sense not just for 'tribal people' but for us in the modern post-industrialised world here and now. Moreover there are now several generations of people in Britain, for example, who have grown up during the time of the psychedelic underground. Their parents took acid back in the day, they themselves did pills and danced until dawn and they are aware that when their kids get older they will probably be doing drugs with long and exotic chemical names. Though under the shadow of illegality the number of people who have accessed the psychedelic state has vastly increased, giving rise to a culture and even family traditions. It's therefore unremarkable that for some people developing their engagement with these experiences in a spiritual sense is something they are inspired to do.

Spirituality is an active attempt to connect with, for want of better words, the sacred. The word 'religion' points to much the same thing, though arguably emphasising a more passive approach to this process. Our (re)connection to the sacred is often imagined as a healing process. On the ground this might look like people breaking additions or entrenched patterns of behaviour, getting more self actualised, becoming more compassionate, and getting into better relationships with themselves, those around them and the planet as a whole. If this is healing then the idea that drugs might help us heal makes total sense.

A second point is that many practitioners of this Western Medicine Way, while having great respect for the native elements of 'designs' such as the peyote ceremony, see these as ongoing, unfolding spiritual systems rather than attempts to ape traditional entheogenic use. Sure when I went to my first Santo Daime ceremony the songs were all in Portuguese but these days songs in English and other languages (including native North American) are part of the repertoire.

British neo-Paganism has grappled by and large successfully in coming to terms with its modern origins (acknowledging that it is inspired by rather than the direct descendant of pre-Christian religion). This willingness to embrace the unfolding of a tradition rather than feeling the need to legitimise practice as old, authentic and/or native has created a richly syncretic culture. This is clearly demonstrated by my last two examples which are attempts to create a ritual 'design' that can serve to create a suitable space for the deployment of two human synthesised substances, one of which (ketamine) is a chemical which as far as we know has only existed on this planet for half a century. In this way modern medicine communities are certainly shamans in the sense that Mircea Eliade uses the term. They are 'technicians of the sacred', adopting designs, tweaking, augmenting and developing their approaches to generating ecstatic and often healing experiences on the part of themselves and their communities.

Some of the rituals I mentioned certainly maintain this future orientation not only by virtue of using new entheogens (MDMA and ketamine) but by creating and managing spaces in which modern technologies may play an important role. This is creating set and setting for the substance using the most appropriate and indeed entertaining tools available. The fast flow of information, even in these often clandestine communities, means that songs, techniques and ideas move around pretty fast. Cultural drift, syncretism and garnering of new techniques and tricks happens so quickly and overtly that claims of a style being set in stone simply don't wash any longer.

Another feature, which is of course the adaptive nature of humans in action, is the fact that the environments being created to 'hold space' are deeply informed by a relationship with the medicine. 'The medicine lets you know what conditions it likes', said one practitioner to me recently. Of course the rave is a large scale version of this process. The drug, MDMA, comes into culture and we create environments designed to maximise its potential. So too in these more explicitly spiritual settings, the psychoactive speaks to those who take it and is an active co-creator of the ceremonial process. But of course, just like evolution, this doesn't create a single type of entity and there are many responses which work for a given time, individual or purpose. Careful attention to the setting of drug experience is a deeply ritualising process and those on the medicine path have a cornucopia of techniques and traditions to inform them. But the spirit of the medicine speaks, and often in a highly animist way. The groups who make use of ketamine, for example, came upon the design described above through a sequence of rituals involving smaller numbers of people, contacting the spirit of ketamine, the Goddess Ketamina and her dolphin familiar Snorky. Through repeated visionary revelations, a ritual design was created suitable for tens of people, keeping the experience sacred, safe and spectacular.

The nature of these groups is in some ways similar to the classical mystery religions or cults of the ancient Greco-Roman world. Secret initiatory processes conducted under the auspices of a particular deity – Mithras, Orpheus, Dionysus or Eleusis. Each of these ancient cults gave the initiate an opportunity to make a special and personal connection to a particular spirit (which, in terms of the ways these things were imagined in the ancient world, is probably a more accurate term than 'god').

This was done through experience, and whether these ancient designs used entheogens is anyone's guess. But certainly in modern day medicine community is an initiate community. There is the act of taking the sacrament in the sacred space and whether you do it once or many times, this is a brave initiatory act. There is no safe word once you've taken that pill. For this reason, and for being brave enough to explore the development of safe, sane and sacred ways of using these powerful medicines in a climate of great fear and hostility, I salute the members of the modern medicine way.

So these are the facts:

- That our culture finds the idea of entheogenic religious experience intelligible
- That these techniques are fluid – they are rapidly shared, adaptable to contemporary conditions and quickly hybridise
- That the medicine itself speaks to participants and they enter a dialogue with it
- That entering this dialogue constitutes a form of initiatory practice

Given that this is the case I submit that the entheogenic genii are well and truly out of the bottle. And that our culture should and indeed must change to accommodate this phenomenon. In practice this must mean changes first and foremost in drug legislation. Now for the record I don't believe that only the religious or spiritual use of drugs should be permitted since there are great recreational as well as more clearly medical possibilities contained in the materials that exist in our society.

It's certainly true that a defence of religious or ethnic use cannot alone oppose the current trajectory of drug legislation (the recent banning of Khat is a good, or rather bad, example of this). However human beings draw much of their personal and social strength from their peak, spiritual and religious experiences. And in this sense, if not necessarily in terms of an argument that claims 'it's my religious right to smoke weed' (or whatever), the spiritual use of sacred drugs is an engine for social change.

The technology, the 'design', of entheogenic ritual is now very accessible in print media, film and the internet. New substances come along all the time and many native and cultivated species are hard if not impossible to ban. Those who walk these medicine paths, as far as they are able, should in my view walk their talk. They are exemplars of the beneficial use of these sacraments and where possible I feel they should act as advocates (though not necessarily evangelists) for these things. As someone who identifies as a magician I know that the world is flexible, it changes and can be changed. For me the emergence of the modern medicine practitioners represents a vital element in a great undertaking. We know that the 'drug problem' is tied closely to big business, criminal cartels, tremendous human suffering, covert government activity and other unpleasant things. For me tackling this issue raises the possibility of bringing, as the medicine community does, heaven and earth together. If we, as a global community, can work to have a better relationship with these physical materials, which themselves can be the basis of the transcendental experience, perhaps we can perfect what the alchemists of old were attempting. We can take the base metal, the horror of the drug war, and transmute it into the gold of a new spirituality which can serve our planet well as we enter the third millennium.

And I pray that the insights, wisdom and power of these medicines can one day be a full and beloved part of our culture.

DYING HIGH: USING PSYCHEDELICS TO MAXIMISE END-OF-LIFE CHOICES BY ENHANCING THE DYING EXPERIENCE

ROBIN MACKENZIE

> *Death is the new sex*
>
> *(Mackenzie, 2009)*

Once upon a time, in the middle of last century, the most pressing question for most young people was will she/won't she, or should we/ shouldn't we. Then came reliable contraception, swiftly followed by new reproductive technologies. Scientific advances enabled us to separate sexual pleasure from reproduction. This disaggregation meant that new questions and a richer vocabulary supplanted old questions. We were suddenly asking how to find the G-spot, become multi-orgasmic and work out what we felt about gay sex. Quite separately, we had to arrive at new moral decisions over having a baby: should it be lawful to buy sperm or eggs off the internet, adopt an embryo, or select between male and female embryos to have a guaranteed boy or girl baby?

How does this apply to death? Currently, choices about death are still in the equivalent of the 1950s. Should we be able to ask for and receive assistance to die if we're very, very ill? Or not? Increasingly, jurisdictions in the West are legalising physician assisted dying. Yet the debate over whether various forms of euthanasia should be lawful has masked a plethora of complex decisions, akin to those following access to reliable contraception and the severing of necessary connections between sex and reproduction. Most choices potentially associated with dying remain relatively unexplored. Do we want to experience dying as a spiritual experience filled with meaning? A blissed out ultimate orgasm? A conscious encounter with the absolute? If so, how do we make this happen? In order to answer these and other questions, we need a commensurate separation between choices over voluntary death and medical treatment, together with specific scientific and technological advances and legal changes. Were death to be demedicalised, so that we had access to means to ensure a peaceful, painless and reliable death without obliging healthcare professionals to act as gatekeepers or to kill us on request, we could explore new ranges of choice over how we might wish to die (Mackenzie, 2009).

A factor fettering choice until now is that we don't know what dying feels like. No-one has come back to tell us. This means that we don't know what our range of choices could be. We need a database of how different ways to die feel. Neuroimaging and neurophenomenological research could make this possible. Brain locations such as the so-called "God spot" (Peterson, 2002) or patterns of activity associated with states of spiritual joy, meaning and compassion, orgasmic bliss and other experiences we might wish to have as we die are increasingly identifiable, so that these states are highly likely to able to be induced by psychedelics (Huynh et al., 2013a; 2013b; Sharp, 2013). Volunteers who agreed to have their brains scanned as they died could help us build up a database of how the subjective experience of dying varies (Mackenzie, 2009). A menu of dying options including the effects of various psychoactive substances and practices like meditation could be created. As with sex, new questions, a richer vocabulary and unprecedented moral decisions would arise.

In order to establish such an infrastructure of choices in dying, ethico-legal and social changes would need to take place. These include the legalisation and regulated supply of currently unlawful recreational psychoactive substances, together with legal reforms providing citizens with access to a lawful, peaceful, painless and reliable death at a time of their choosing. As the arguments for and against these measures are well known, I shall not repeat them. Rather, in the rest of this piece I argue in favour of allied changes promoting using psychedelics to maximise end-of-life choices by enhancing the dying experience.

Baby-boomers, the generation born after the war who were young in the 1960s and 1970s but are now ageing, are crucial. Their increased life expectancy is matched by a developed sense of entitlement, autonomy and economic clout, which will lead them to demand control over not only when but how they die. This pressure, combined with their relaxed attitudes to recreational drug use, will strengthen contemporary movements supporting assisted dying, the demedicalisation of death and reform of the drug laws to promote cognitive liberty. Opening up choices in death to include enhancing the dying process with psychedelics dovetails with established cultural practices whereby Westerners routinely take lawful, unlawful and prescription mind-altering drugs to enhance their moods (Kupelian, 2014), and commodify Eastern spiritual practices such as meditation, mindfulness and yoga while co-opting them into clinical practice.

Evidence from the neuroscience of the range of dying experiences potentially available could focus and underpin such sociocultural vectors. Constructed taxonomies of the neurophenomenology of psychedelics at end-of-life according to health status and individual characteristics, with an acknowledged place for psychedelics in palliative care, bereavement services, self-chosen death and cognitive enhancement, are an essential prerequisite to maximise end-of-life choices by using psychedelics to enhance the dying process.

USING PSYCHEDELICS TO ENHANCE DEATH: NEUROSCIENCE, BABY-BOOMERS, THE SECULARISATION OF SPIRITUALITY AND THE COMING COMMODIFICATION OF THE DYING PROCESS

Western societies' link between religiously inspired unease over pleasure and prohibition of recreational psychoactive substance use (Mackenzie 2010; 2008; 2006) is disintegrating, partly because the war on drugs has failed, and partly with the incorporation of Eastern spiritual practices, particularly in the form of mindfulness meditation, into Western clinical practice and commercial enterprise. Since the secular forms of Eastern spiritual practices provide pleasure in the form of positive emotions (Vaillant, 2014), the moral approbation traditionally accorded recreational psychoactive substance use loses its force as pleasure is revealed as good for us (Mackenzie, 2014). While concerns over safety remain, most may be addressed via legalisation and regulation of supply and access. An increasing number of jurisdictions are taking this step, such as the state of Washington, as the clinical and commercial potential of hitherto unlawful psychoactive substances becomes clearer (Balter, 2014; La Ganga 2013; Svaldi, 2014; Taborek, Bost & Gammelhoft, 2014). Although addiction remains a problem, it is noteworthy that psychedelics are non-addictive; indeed some of them have shown promise in treating addictions (Mackenzie, 2014).

A significant factor in altering perspectives is evidence from neuroscience, which has enabled a more detailed understanding of how exogenous and endogenous psychoactive substances operate in human minds, brains, bodies and behaviour. The revelation of homologies between psychedelics and spiritual practices like mindfulness meditation is particularly crucial, since it suggests that in appropriate circumstances they may achieve very similar therapeutic and pleasurable ends. Evidence indicates that using mindfulness meditation we could potentially get to feel healthier, function more effectively and feel happier, and that we could also do so through taking psychedelics safely, without its taking so long to get the same effects. Such speedier outcomes appeal to healthcare providers and commercial enterprises, as well as consumers, particularly the baby-boomers. I shall now elaborate on the implications of these points.

EVIDENCE FROM NEUROSCIENCE

Debates on the moral complexities of choice over how we die will be informed by the neuroscience of the dying process, the neurophenomenology of psychedelics use, particularly at end-of-life, and driven by existing social forces including the baby-boomers' assumption of choice over death as well as in life and the recognition of the clinical and commercial potential of spiritual experiences induced by legalised psychedelic substances.

While neuroscience has undoubtedly given rise to a degree of neurohype, the use of neuroimaging to evidence different states of the mind and brain has proven clinically and commercially profitable. Meditation provides an illuminating example of this. Neuroimaging has fostered investigations of the potential of meditation, which includes arresting cognitive decline (Gard, Holzel & Lazar, 2014; Luders, 2014), reducing pain (Grant, 2014), improving cognition and memory in neurodegeneration (Newberg et al., 2014), enriching paradigms and maps of the neural system (Brewer & Garrison, 2014; Loizzo, 2014), improving healthy brain functioning generally (Britton et al., 2014), improving conscious regulatory control over emotion and attention throughout the life span (Tang, Posner & Rothbart, 2014) and enabling access to a variety of spiritual Alternative States of Consciousness (ASCs) or Non-Ordinary States of Consciousness (NOSCs), (Josipovic, 2014; Travis, 2014; Vago, 2014).

Neuroimaging makes it possible to demonstrate in an accessible fashion how meditation could be drawn upon to address current issues in healthcare provision, through improving cognitive and emotional functioning. Neuroimaging studies of meditators suggest that the ability to move between ASC/NOSCs by choice through meditation would be beneficial and achievable (Josipovic, 2014; Travis, 2014; Vago, 2014). The ability to move intentionally between ASC/NOSCs in an informed fashion could thus improve human health and flourishing, as well as providing us with increased options over how we die (Mackenzie, 2014). Placed in the context of contemporary healthcare provision, this renders it likely that tools which enable choices allowing movement between ASC/NOSCs would prove popular and so will become commodified, as have Eastern spiritual practices like mindfulness meditation. Moreover, neuroscience has illuminated psychedelics' potential to function as tools to enable us to move between states of mind including ASC/NOSCs by choice (Mackenzie, 2014). While psychedelics nonetheless remain outlawed psychoactive substances, it appears increasingly likely that their commercial potential in the clinical treatment and self-help industries will lead to legalisation with the regulated supply chains necessary to ensure safe consumption. The investment being poured into the legalised cannabis supply industry in the United States at present is one example of the probable outcome (Balter, 2014; La Ganga, 2013; Svaldi, 2014, Taborek, Bost & Gammelhoft, 2014).

CLINICAL BENEFITS AND COMMODIFICATION OF SPIRITUAL EXPERIENCE

The imprimatur of neuroscience, popularised in political announcements for the funding of various years and decades of the brain, has anchored and provoked a variety of societal preoccupations. One is the Western world's established incorporation of Eastern spiritual practices, in the form of yoga, meditation and mindfulness into clinical practice, initially as techniques to access spiritual ASC/NOSCs for stress-relief. A prime example of this is Jon Kabat-Zinn's treatment of chronic pain in patients in the 1970s with a secular programme he called Mindfulness Based Stress Reduction (Kabat-Zinn, 2008). This initiative has spear-headed the acceptance of mindfulness into mainstream institutions in healthcare, medicine, psychology and hospitals, as well as more broadly into schools, corporations, the legal profession, prisons and sports coaching. Health services in nation states have accepted mindfulness and meditation as complementary practices to address chronic conditions such as pain, mental illnesses and age-related declines in physical and mental health (for more specific information on clinical guidelines in the UK, see www.nice.org.uk). Clinical psychologists have adopted mindfulness and meditation as staples in their practice (Gilbert, 2013).

This incorporation of Eastern spiritual practices into Western clinical treatment and the self-help industry represents a secularisation of spirituality with a variety of salient consequences. Secular spirituality becomes severed from the culturally problematic relationship with pleasure associated with the Western version of Abrahamic or Judeo-Christian religions. Positive emotions involving pleasure can become associated with spirituality, while neuroscience demonstrates commonalities between pleasurable psychoactive substance use and spiritual experiences. This calls into question legislation prohibiting psychoactive substance use motivated by cultural and commercial shaping of what constitutes acceptable experiences of pleasure (Mackenzie, 2010; 2008; 2006). Moreover, demonstrable health benefits of spiritual experiences and practices are manifold (Mackenzie, 2014; 2009; 2006).

BABY-BOOMERS, AGEING AND END-OF-LIFE DECISIONS

Neuroimaging has fostered a refocusing of practices like meditation as tools to treat ageing brains as life expectancies rise, demographic changes entail higher proportions of the elderly in Western populations, and cognitive changes become able to be monitored. Exercise regimes, memory clinics, sudoku and mindfulness apps promise to halt or reverse the slide of our ageing brains into cognitive decline, dementia and death. In the same way that profits have been generated by our media-induced desires to look a certain way, forever young, the alternative health anti-dementia/anti-Alzheimer's disease industry has generated significant profits by defining the characteristics of the ageing brain as disease or disorder, to be addressed through purchasing various forms of treatment as self-help measures.

This has impacted particularly on baby-boomers, who have become the well-off worried well, as they seek to prevent not only the personal decline they see or have seen in their parents, but also the unenticing prospect of dying dependent and demented. They are thus are uniquely motivated to seek out means to experience the benefits of meditation, and expect to be able to buy them. Baby-boomers are accustomed to exercising choice. They also often have the financial means to do so.

This means that they are a significant sector of the economic public, with the voting and spending power to influence policies and national or international economies.

Meditation and mindfulness have a particular appeal for them as they hold the promise of allowing choice between different states of mind and brain, as well as of postponing or reversing the ageing process. However, most research within neuroscience into the marked effects of meditation have been carried out upon experienced practitioners associated with particular belief systems who have meditated for several hours daily over years. Baby-boomers anticipating age-related decline are unlikely to be willing or able to put in this degree of time and self-discipline to achieve a means of moving between ASC/NOSCs at will. As research suggests that psychedelics offer a means of doing so, and many of not most of the baby-boomers have experienced recreational drug use, pressure to reform the drug legislation accordingly to allow access to safe, lawful supplies of psychedelics is likely.

THE NEED FOR FURTHER EVIDENCE

There is increasing evidence that health and spiritual benefits associated with meditation and psychedelics are commensurate (Mackenzie, 2104). In appropriate circumstances psychedelics provide transcendent spiritual experiences associated with consciousness expansion, life-altering experiences of subjective meaning and enhanced health (Mackenzie, 2013; 2014). Arguing that psychedelics may be conceptualised as potentially choice-enhancing catalysts of consciousness, King describes consciousness expansion as 'epilogenic consciousness', a state where ordinarily autonomic decision-making processes which are consciously attended to may enable us to choose between them (King, 2013). The neurochemistry of the ASC/NOSCs to which psychedelics give rise provide potential mechanisms demonstrating how psychedelics may enable us to choose between them.

Vaillant suggests that spirituality is made up of eight positive emotions: awe, love/attachment, trust/faith, compassion, gratitude, forgiveness, joy and hope, which psychiatric treatment should focus on (Vaillant, 2013). He believes that spirituality is all about emotion and social connection, which are more dependent on the limbic system rather than the cortex. This position, with an implied crucial role for endogenous psychedelics, is supported by research into the neuroscience of sleep and near death experiences (NDEs). Nelson argues that:

> *"We have strong indications that much of our spirituality arises from arousal, limbic and reward systems that evolved long before structures made the brain capable of language and reasoning. Neurologically, mystical feelings may not be so much beyond language as before language"* (Nelson 2012)

Further ethically grounded research into the circumstances under which humans produce endogenous entheogens, such as the neurochemistry of the role of DMT in the dying process (Mackenzie, 2009; Strassman, 2001; Wutzler, 2011) is essential if the conflict between neuroreductionism and non-falsifiable assertions of seemingly paranormal events or religious/mystical experiences is to be resolved (Mackenzie, 2014). Current research into the neuroscience of the dying process suggests that dying itself, as opposed to NDEs, can produce experiences which resemble those associated with psychedelics and NDEs (Borjigan et al., 2013; Mobbs & Watt, 2011). However, research indicates that the pain-killing and sedative drugs associated with current end-of-life palliative care practices make it less likely that NDEs will take place (Satori, 2014). This implies that few of us achieve an optimum dying process. Achieving the crucial legal changes necessary to ameliorate this depends partly on the ability to produce research to anchor evidence-based law and policy.

CONCLUSION

The neuroscience of how psychedelics and ASC/NOSCs might contribute to human health thus holds the potential to transform how we live and flourish and how we die (Mackenzie, 2014). Official permission to research unlawful psychedelics/psychoactive substances has led to their becoming increasingly accepted as having therapeutic properties, like Ecstasy/MDMA for PSTD, ibogaine for addiction and ketamine for depression: up-to-date information may be found at the Multidisciplinary Association for Psychedelic Studies (MAPS) website, www.maps.org. For example, at John Hopkins, psilocybin has been shown to provide spiritual meaning and pain relief for terminal cancer patients (Grob, Bossis & Griffiths, 2103). This suggests that taking psychedelics in a controlled context, accompanied by an experienced guide, should guard against the 'bad trips' those taking potentially adulterated unlawful psychoactive substances have sometimes experienced. Hence taking psychedelics as part of dying, with a midwife of death or psychopomp to show the way, could become part of palliative care, or an exercise in cognitive liberty (Mackenzie, 2014; 2009). Moreover, those suffering existential angst after the loss of a loved one might also benefit from post-bereavement practices incorporating psilocybin.

How likely is this? Although the baby-boomers are ageing, they have no wish to be old, nor to give up their lifelong assumption of a continuing entitlement to feeling good and to exercising autonomous choice. Moreover, spirituality is increasingly coded as a secular means of enhancing health and well-being rather than as affiliated with a religion. Hence the commodification of spiritual experiences which make us feel good, enabled by advances in neuroscience up to and including the experience of dying, is increasingly plausible. Consider NDEs. Research into the neuroscience of REM sleep and dreams suggests that many of the apparently transcendent spiritual experiences reported by those who have had an NDE may arise from disjunctions of mechanisms which normally provide smooth transitions between different states of consciousness. Looking down on one's body, engaging in a review of one's life, proceeding towards a welcoming bright light at the end of a tunnel and feeling a mystical union with the universe may soon be able to be reliably induced through reverse engineering for their beneficial personal consequences. These include increased altruism, empathy and well-being, together with the loss of the fear of death. Endogenous psychedelic substances produced in our bodies, like DMT, or the 'God molecule', may provide similar effects to those of exogenous psychedelics like psilocybin. As the subjective value of spiritual experiences in end-of-life contexts is increasingly acknowledged, secularised and commodified, a reframing of the good death and of psychedelics is ushered in. Marketed as preserving autonomy and maximising a spread of good time options, this would surely ease the process whereby our future preoccupations move from G-spot to God spot.

Balter, J. (2014). *Marijuana is legal: time to regulate and tax it.* 8 January. http://www.bloomberg.com/news

Borjigan J. et al. [10 authors] (2013). Surge of neurophysiological coherence and connectivity in the dying brain. *Proceedings of the National Academy of Sciences* 2013; doi 10.1073/pnas.1308285110.

Brewer, J. & Garrison, K. (2014). 'The posterior cingulate cortex as a plausible mechanistic target of meditation: findings from neuroimaging' *Annals of the New York Academy of Sciences,* vol 1307, pp. 19-27.

Britton, W. et al. (2014). 'Awakening is not a metaphor: the effects of Buddhist meditation practices on basic wakefulness' *Annals of the New York Academy of Sciences,* vol 1307, pp. 64-81.

Gilbert, P. (2013). *Mindful compassion.* Constable & Robinson, London.

Gard, T., Holzel, B. & Lazar, S. (2014). 'The potential effects of meditation on age-related cognitive decline: a systematic review' *Annals of the New York Academy of Sciences,* vol 1307, pp. 89-103.

Grant, J. (2014). 'Meditative analgesia: the current state of the field' *Annals of the New York Academy of Sciences,* vol 1307, pp. 55-63.

Grob C., Bossis A. & Griffiths R. (2013). Use of the classic hallucinogen psilocybin for treatment of existential distress associated with cancer. In Carr, B. & Steel, J., (eds.) *Psychological aspects of cancer: a guide to emotional and psychological consequences of cancer, their causes and treatment.* 291-308. Springer, New York.

Huynh, H., Willemsen, T. & Holstege, G. (2013). Female orgasm but not male ejaculation activates the pituitary. *Neuroimage* vol 76; 178-182.

Huynh, H., Willemsen, A., Lovick, T. & Holstege, G. (2013). Pontine control of ejaculation and female orgasm. *Journal of Sexual Medicine* vol 10; 3038-3048.

Josipovic, Z. (2014). 'Neural correlates of nondual awareness in meditation' *Annals of the New York Academy of Sciences,* vol 1307; 9-18.

Kabat-Zinn, J. (2008). Mindfulness based interventions in context: past, present, future. *Clinical Psychology Science & Practice* vol 10; 144-156.

Kerr, C., Sacchet, M., Lazr, S., Moore, C. & Jones, S. (2013). 'Mindfulness starts with the body: somatosensory attention and top-down modulation of cortical alpha rhythms in mindfulness meditation. *Frontiers in Human Neuroscience,* doi: 10.3389/fnhum.2013.00012 .

King, D. (2013). *'A dose by yet another name'.* UKC Psychedelics Society, Canterbury, 19 November.

La Ganga, M. (2013). Washington state reveals recreational pot rules. *Los Angeles Times*. 5 September. http://articles.latimes.com

Kupelian, D. (2014). *'Seventy million Americans take mind altering drugs'*. www.wnd.com , online 9 February 2014, accessed 19 February 2014.

Loizzo, J. (2014). 'Meditation research, past, present and future: perspectives from the Nalanda contemplative tradition'. *Annals of the New York Academy of Sciences*. Vol 1307; 43-54.

Luders, E. (2014). 'Exploring age-related brain degeneration in meditation practitioners'. *Annals of the New York Academy of Sciences*. Vol 1307; 82-88.

Mackenzie, R. (2014). 'What can neuroscience tell us about the potential of psychedelics in healthcare? How the neurophenomenology of psychedelics research could help us to flourish throughout our lives, as well as to enhance our dying' *Current Drug Abuse Reviews*. Forthcoming.

Mackenzie, R. (2013). *Entheogens, society and law: towards a politics of consciousness, autonomy and responsibility: foreword*. Waterman, D. & Hardison, C. W. (eds.) Melrose Press, Ely.

Mackenzie, R. (2010). The neuroethics of pleasure and addiction in public health strategies: moving beyond harm reduction: funding the creation of non-addictive drugs and taxonomies of pleasure'. *Neuroethics* vol 4(2); 103-117.

Mackenzie, R. (2009). Reframing the good death: enhancing choice in dying, neuroscience, end of life research and the potential of psychedelics in palliative care. In M. Freeman and O. Goodenough, eds., 2009. London: Ashgate; 239-263.

Mackenzie, R. (2008). Feeling good: the ethopolitics of pleasure, psychoactive substance use and public health and criminal justice system governance: therapeutic jurisprudence and the drug courts in the United States of America'. *Social & Legal Studies* vol 17(4); 513-533.

Mackenzie, R., (2006). Addiction in public health and criminal justice system governance: neuroscience, enhancement and happiness research. *Genomics, Society and Policy* vol 2(2); 92-109.

Mobbs, D. & Watts, C. (2011). There is nothing paranormal about near death experiences: how neuroscience can explain seeing bright light, meeting the dead or being convinced that you're one of them. *Trends in Cognitive Science* 15(10); 447-449.

Nelson, K. (2011). *The god impulse: is religion hardwired into our brains?* Simon & Schuster, London.

Newberg, A. et al. (2014). 'Meditation and neurodegenerative diseases' *Annals of the New York Academy of Sciences*. Vol 1307; 112-123.

Peterson, G. (2002). Mysterium tremendum. *Zygon*. Vol 37; 237-254.

Satori P. (2014). *The wisdom of near-death experiences*. Watkins, London.

Sharp, P. (2013). Meditation induced bliss viewed as a release from conditioned neural (thought) patterns that block reward signals in the brain pleasure centre. *Religion, Brain & Behaviour*. http://dx.doi.org/10.11080/2153599X.2013.826717.

Strassman, R. (2001). DMT spirit molecule: a doctor's revolutionary research into the biology of near-death and mystical experiences. Park St. Press, Rochester.

Svaldi, A. (2014). High Times launches private equity fund for marijuana investment. *Denver Post*. http://www.denverpost.com

Taborek, N., Bost, C. & Gammeltoft, N. (2014). Pot shares rally 21% to 1700% as speculators see green. *Bloomberg.* 9 January http://www.bloomberg.com

Tang, Y., Posner, M. & Rothbart, M. (2014). Meditation improves self-regulation over the life-span'. *Annals of the New York Academy of Sciences,* vol 1307; 104-111.

Travis, F. (2014). 'Transcendental experiences during meditation'. *Annals of the New York Academy of Sciences.* 1307; 1-8.

Vago, D. (2014). 'Mapping modalities of self-awareness in mindfulness practice: a potential mechanism for clarifying habits of mind' *Annals of the New York Academy of Sciences.* 1307; 28-42.

Vaillant, G. (2013). Psychiatry, religion, positive emotions and spirituality. *Asian Journal of Psychiatry* 6; 590-594.

Wutzler A., Mavrogiogou P., Winter C. & Juckel G. (2011). Elevation of brain serotonin during dying. *Neuroscience Letters 2011.* 498; 20 -21.

PSYCHEDELIA AND VISIONARY ART – THE WORK OF ART AS THE RESULT OF INTERACTION BETWEEN CULTURE AND NON-ORDINARY STATES OF CONSCIOUSNESS

JOSÉ ELIÉZER MIKOSZ

This essay investigates the poetics of visionary art as a result of interactions between local cultures and non-ordinary states of consciousness. We seek, through brief examples, to draw a parallel between images produced in the past, images produced by Amazonian tribes, and the production of contemporary artists and to point out series of meaningful coincidences between them. Despite local cultural differences, it is possible to observe similar patterns of *non-ordinary states of consciousness* in the context of artistic works; at the same time, such an approach also brings up a whole historical background to the current – and so-called – visionary art.

RELIGIOUS EXPERIENCE AND MYTHS

Experiences of entering into non-ordinary states of consciousness can take us to a world different from the ordinary, material and rational world of our everyday life, therefore many times associated with a 'spiritual realm'. Some advocate the real existence of such a realm; others, however, may consider that it is nothing more than characteristic reactions of our nervous system. It is possible to divide this type of experience into two distinctive forms. The first consists in direct contact of the individual with this spiritual realm. This is probably the case of many avatars and prophets since the ancient ages, religious founders, priests and/ or shamans.

The other form is the one experienced by disciples. They 'believe' in those spiritual leaders, not merely due to their innocence and unpreparedness, fears and doubts in the face of a mysterious life they are not able to explain, but also because, somehow, they feel within themselves a resonance of what is conveyed by those masters, thus giving them confidence and coherence as stated by Campbell:

> *"Any person who engages with the work of literary creation knows that we open up ourselves, it is a kind of surrender, and the book talks to us and builds itself. To some extent, you become the conveyer of something that was transmitted to you from what is called the Muses, or, in biblical terms, 'God'. This is not a manner of speaking, this is a fact. Once the inspiration results from the unconsciousness, and once the mind of the people of any small society has much in common with what the unconsciousness is concerned, what the shaman or the prophet brings to light is something that exists latent in anyone, just waiting to be brought to light."* (Campbell, 1991)

It is very likely that certain myths, for instance, some indigenous legends, have emerged from visions received in non-ordinary states of consciousness. However, in time, they may suffer alterations and adaptations according to transformations of time and local culture, deviating them, apparently, from the initial associations, or even being substituted by *symbolic synonyms*, in other words, different images with the same meaning:

> *"Well, automobiles entered mythology. They entered the dreams. And aircrafts are long in service of imagination. The flight of the aircraft, for instance, acts in the imagination as a liberation from earth. It is the same thing with birds, in a certain way. The bird is a symbol of the liberation of the spirit in relation to its earth's imprisonment, in the same way that the serpent symbolizes the earth's imprisonment. The aircraft plays this part nowadays."* (Campbell, 1991)

A myth, as a sacred account about creation, told by a people, can sound funny to others: what is sacred for some people, is superstition for others, but this does not reduce the power and efficiency of the myth in its place of birth (Lewis-Williams & Pearce, 2005). Myths are inserted in a society, people inherit them and assimilate the culture in which they were born. People would feel bad and ridiculous should they be obliged to dress, or undress, according to practices and customs of a culture different from their own. Similarly, the individual absorbs the religion available around them, becoming a Christian, Jew, Muslim, Hindu, or Huichol.

A deep study of these characteristics is something very complex, but it is possible to find some examples to illustrate the question. To begin with, we will study some characteristics of the human mind while in non-ordinary states of consciousness, and, in parallel, we will show some attempts to make visual representations of these experiences by some individuals.

SHAMANISM AND NON-ORDINARY STATES OF CONSCIOUSNESS

What would be the main characteristic one could notice in spiritual phenomena? The answer is that they seem to be linked to the human ability of changing their states of consciousness through what we will thus call *shamanic-spiritual techniques* or *psychedelic techniques*. Many of them are well-known and used by various religions: meditation, fasting, celibacy, privation of senses, stimulation of the cerebral visual cortex by stroboscopic lights, yoga, tantric, usage of plants and psychoactive substances, chanting of mantras and prayers, physical exhaustion, sacred dances, usage of specific music that ranges from shamanic drums and maracas, to modern electronic music. The non-ordinary states of consciousness cover a series of characteristics, such as the contact with different levels of reality that are beyond both rational process and material world; however, what precisely interests us in this study are the visual phenomena that some of those techniques can facilitate, in other words, what those techniques can cause in terms of visual stimuli – generally called, in a reductive way, hallucinations – and the production of visual arts as a result of those experiences.

Huxley (1957) observes that the brain seems to work as a reducing valve of perception, a filter for reality, "to make possible biological survival" (Huxley, 2004). Our consciousness receives a torrent of impressions from the outer world and our own body that needs to be filtered. In fact, we are not simultaneously aware of all the stimuli coming from our senses. What comes to our consciousness must, as it were, follow a hierarchy of priorities.

In the same way that the outer world impressions are filtered, the inner impressions, in other words, the memories, thoughts and unconscious contents, must go through some kind of selection. In 'normal conditions', people relate to the world through this filtering. However, *shamanic-psychedelic-spiritual techniques* can avoid this filtering, and, in some cases, even intensify the brain reception of certain impressions coming from the outer or inner world. Hancock, about some probable brain characteristics, says:

> *"Theoretically the brain could be as much a receiver as a generator of consciousness, and thus might be fine-tuned in altered states to pick up wavelengths that are normally not accessible to us."* (Hancock, 2007)

The shamanic practices are associated with the non-ordinary states of consciousness and are very old and similar among people of different parts of the world, a reason why some researchers use the expression *shamanic state of consciousness* (Harner, 1982). In this state it is possible to observe the spectrum of consciousness divided into three stages (Lewis-Williams, 2004):

Stage 1, of *entoptic phenomena*, visual phenomena that occur between the eye and the cortex, independently of the material world, but that can be projected on outer world objects. These patterns usually consist of a variety of geometric, colourful and luminous forms in a fractal combination. Klüver divides the entoptic phenomena into four categories, also called *form constants* or *phosphenes*: "(i) gratings, lattices, fretworks, filigrees, honeycombs and chequer-boards, (ii) cobwebs, (iii) tunnels, funnels, alleys, cones and vessels, and (iv) spirals" (Klüver, 1966). Lewis-Williams identifies seven types of categories:

- a grid and its development into a lattice or expanding hexagon pattern
- set of parallel lines
- bright dots and short flecks
- zigzag lines, reported by some subjects as angular, by others as undulating
- nested catenary curves, the outer arc of which comprises flickering zigzags (well known to migraine sufferers as the 'fortification illusion')
- filigrees, or thin meandering lines
- spirals (vortex), we can identify these patterns since rupestrian art

Stage 2, or *construal*, which can be understood as a process of interpretative construction: in this stage, the individual tries to make sense out of the entoptic forms – similar to when one observes undefined or ambiguous images, such as spots, cloud formations, folds of fabric, which can be transformed into familiar shapes, such as animals, people, faces, etc. according to propensity, cultural aspects, and many momentary influences. For example, when we look at the sequence of colon, hyphen, and parenthesis and we see a smiling face. By getting closer to stage 3, the experience with vortex or tunnels, with shining light on the background, is common and it is often associated with the near-death experience. It is at this point that:

"... many individuals report experiences with vortex or a spinning tunnel that seems to encircle and attract them to its bottom"

(Horowitz, 1975)

Tunnels seem to be related to certain brain structures; Bressloff et al., (2001) describe a mathematic investigation about the possible origin of these images, assuming that the connection patterns between the retina and the striate cortex (V1) – the retino-cortical pathway – and the neural circuits in V1, local and lateral, are what determine these geometries.

Finally, **Stage 3**, the 'hallucinations', in other words, visions in which more complex scenes are formed. The individual is not always able to draw a distinction between their experience and the material world. Complete visions of all kinds can be formed, bizarre somatic sensations can happen, such as physical deformations, transfigurations into animals and plants, or others.

This division in stages does not mean that people necessarily go through all of them, nor that the passage from one stage to another has rigid boundaries – on the contrary, the passage happens gradually. Cultural information can influence the individual's expectations and interests, and accentuate one stage or another.

VISIONARY ART

"Visionary Art is an art where the production is subject to the resulting experiences of non-ordinary states of consciousness"

(Mikosz, 2009)

The visions experienced in non-ordinary states of consciousness are usually perceived as genuine by individuals, even by those who are not shamans. They are experiences of 'another world', which can only be translated into 'this world' in a descriptive and symbolic way, mediated by each individual's culture and internal repertoire. This is a fundamental point, for it demands a profound reflection on how we understand the foundations of religious faith. Depending on cultural influences, language, expectations and personal preferences, the individual's emotional state, environmental circumstances, in short, contingent elements known under the expression *set and setting*, the experience can be guided as to what will be experienced or seen, respecting certain limits of possibility. For instance, the vision of a spiral can acquire depth as a vortex or a tunnel. The images can also be seen multiplied (poliopsia) and/or in integration with other objects in the scene, as happens in the geometric drawings of people and animals, arisen from visions, which the Tukano people perform on their canoes (Lewis-Williams & Pearce, 2005). The vision of spirals and their various associations in each culture, such as snakes, labyrinths, circles, tunnels, ladders, follow similar principles of transformation (Mikosz, 2009). It should be noted here that, even though this paper is mainly focused on visual phenomena, various sensations, not only visual ones, are present in the non-ordinary states of consciousness.

In the case of the cosmological experience...

> *"...the shamanic technique par excellence consists in the passage of one cosmic region to another, from Earth to Heaven or from Earth to Hell".*
> *(Eliade, 2002)*

Shamans perform this through a structure that is part of the universe and that connects this 'layered cosmos': the *axis mundi*, the axis of the world, or yet, the cosmic pillar, which goes through a 'doorway', a 'hole', from where the gods, the dead, the shamans can ascend and descend on their celestial or infernal trips (ibid.).

The following images show some examples of the aforementioned entoptic elements, both in representations of indigenous peoples of the Amazon forest (Tukano and Shipibo) as well as in the work of two artists from Western art history (Bosch and Blake), and of some visionary contemporary artists that the author had access to.

The Barasánas (Tukano family), indigenous people who use the ayahuasca in their rites, refer to images of vertical wavy parallel lines

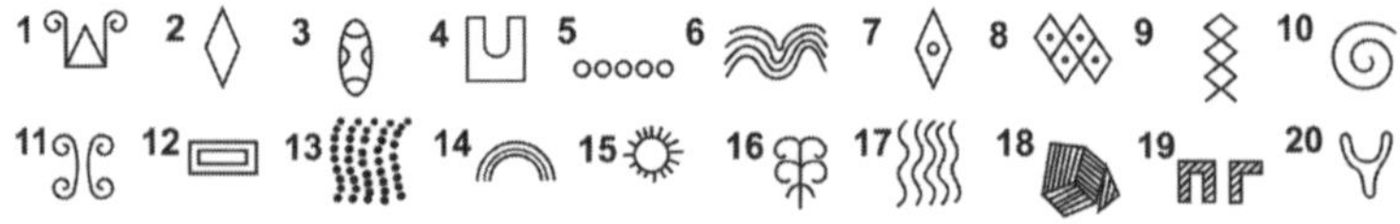
1 2 3 4 5 6 7 8 9 10
11 12 13 14 15 16 17 18 19 20

(17th pattern in the chart above) which are part of *Type 4* of the seven most recurrent shapes previously mentioned – undulating zigzags – as symbolising "...the creative thought and, sometimes, the energy of the solar creator itself". An arch, 14th pattern, with many colourful parallel lines of *Type 5*, can symbolise the rainbow and, in one of the interpretations, the father-sun's penis. The San people from South Africa enter a state of trance through dances. They concentrate on bright lines of *Types 2* and *6*, which they believe to be *filaments of light* by which they climb, or along which they fluctuate towards the Great God in heaven (Lewis-Williams & Pearce, 2005). The 10th pattern, a spiral of *Type 7*, represents incest and forbidden women.

The painting *Ascent of the Blessed*, by Hieronymus Bosch, depicts a tunnel the souls pass through on their way to Heaven. Individuals who have undergone near death experiences report similar images of tunnels with a blue light irradiation at the end.

Another way of representing these passages are the images of stairs that go from earth to the sky like in the painting *Jacob's Dream* by William Blake, based on the description of Jacob's vision in the

FIGURE 1: Codified drawings of the Barasána people in Brazil. Drawing by the author based on Reichel-Domatoff (1978)

FIGURE 2: Ascent of the Blessed – in detail. Hieronymus Bosch (1450-1516)

Bible, where angelical beings go up and down these stairs. In a different and syncretic culture between shamanic elements and Christianism, the same idea can be represented differently like in the painting *Spirits Descending on a Banco,* by Peruvian painter Pablo Amaringo, where spirits go up and down not using stairs, but by way of a blue light spiral, intermediated by the figure of a Shaman (Banco). In the

FIGURE 3 & 4: Jacob's Dream. William Blake (1757-1827), Spirits Descending on a Banco (medium). Pablo Amaringo (1938-2009)

Amazon forest, the trees often represent a symbol of the *axis mundi* for the peoples who live there, since stairs are not common in that environment and therefore are not part of the peoples' imagination like in other cultures.

Tunnels, stairways, ziggurats, towers (like the Babel Tower described in the Bible, built with the purpose of reaching Heaven), just like mountains, represent this ascent toward the spiritual realm. The image shown right, by artist Daniel Mirante, shows this sense of ascent and of the sacred in the painting *Song of Vajra*.

The Shipibo-Conibo Indians, native to the Peruvian Amazon, work with geometrical patterns that include catenary curves (*Type 5*) that end in little spirals (*Type 7*), zigzags (*Type 4*) that create a continuous background pattern like filigree (*Type 6*). These patterns are embroidered on a variety of fabrics, appear on pottery, body painting and ceremonial garments, among others. They may be simply drawn with black ink or in several colours, with small changes in the resulting patterns.

A similar pattern was used by the visionary artist,

FIGURE 5: Song of Vajra. Daniel Mirante

FIGURE 6: Shipibo-Conibo pattern

professor, and writer Laurence Caruana, the current director of the Vienna Academy of Visionary Art. In the painting *Vine of the Dead*, Caruana used a similar pattern as a background for the image of Christ; he comments:

> *"Then, during an all-night ceremony led by Native peoples, I ingested ayahuasca for the first time. Over the course of six hours, I reviewed my life, confronted my own death, and experienced a kind of personal Last Judgment. Toward early morning, this personal vision-quest transformed into a more archetypal vision. Wherever I turned my eyes, I saw the sacred patterns which constitute (so it seemed to me) the interconnected space and substance of our very souls. Weeks later, I was amazed to discover that the Shipibo-Conibo tribes of the Amazon, who treat ayahuasca as a sacrament, sew these same patterns onto all their ceremonial vestments, recognizing them as 'patterns of the soul'"*
>
> (Caruana, 2006)

Parallel lines (*Type* 2), zigzag undulating lines (*Type* 4), and nested catenary curves (*Type* 5) can be found in many visionary works like Amanda Sage's paintings. *Limbic Resonance*, overleaf, shows two figures

FIGURE 7: The Vine of the Dead. Laurence Caruana

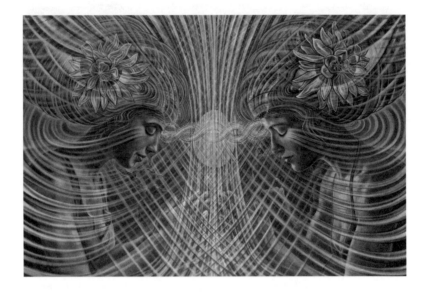

in synchrony and communion amidst vibrations and energy exchange symbolised by luminous lines conveying the idea of an harmonious-amorous whole, states also associated with forms of spiritual communion typical of certain non-ordinary states of consciousness.

In many of the works by artist Andrew Gonzales, spirals and vortexes (*Type* 7) are present, usually transformed or merged with feminine figures. As previously seen, the spirals may represent the idea of passage, just as labyrinths, mandalas, tunnels and stairs that, in fact, are not just images seen during visions, but may also be sensorially experienced in several ways by an individual during non-ordinary states of consciousness. Another very common element associated with spirals is water. Actually, the association of water as a primordial element of creation is present since ancient Egypt. It is common to find spirals as a symbol of creation, water and lunar fecundity, associated with the feminine, the mother, the sea (Chevalier & Gheerbrant, 1999), all harmoniously present in the work of the artist.

FIGURE 8: Limbic Resonance. Amanda Sage

FINAL WORDS

We have discussed the visual coincidences arising from non-ordinary states of consciousness, emphasising that the experiences may produce similar images in individuals all over the world; conversely, they may undergo variations and adaptations owing to the environment and local culture the individuals are immersed in.

This paper attempted, from the vast quantity of examples that would not fit here, to present a small sample of visual representations from the remote past and others by contemporary artists that are both illustrative and result from *non-ordinary states of consciousness.* In the same way that indigenous people can consider some geometrical patterns to be sacred, because they were 'seen' through experiences of contact with other levels of reality, the 'spiritual realm', science tries to investigate those religious phenomena not as merely hallucinatory reveries, but as a result of legitimate experiences, regardless of any evidence of the existence of such realms. Nowadays, the very nature of hallucination is being questioned, considering that because they do not exist in the material objective world, it does not mean that they cannot be part of a perfectly normal human psyche, thus presenting recurrent characteristics that are measurable by science. The visual representations, examined in an interdisciplinary way with human and biological sciences, help us to understand some evidence regarding the choice of certain geometric patterns used in the representations, as well as their modification and sophistication in more complex societies.

FIGURE 9: Spiral Reverie. Andrew Gonzales

Bressloff, P.C., Cowan, J.D., Golubitsky, M., Thomas, P.J. and Wiener, M.C. (2001) Geometric Visual Hallucinations: Euclidean symmetry and the functional architecture of striate cortex. Philosophical Transactions of the Royal Society B: Biological Sciences. 356 (1407). p.299-330.

Campbell, J. (1991) *O Poder do Mito.* São Paulo: Palas Athena.

Caruana, L. (2006) *I Am the True Vine the Ayahuasca the Vine of the Dead.* [Online] Available from: http://www.lcaruana.com/webtext/mort.html. [Accessed: 18ᵗʰ June 2014].

Chevalier, J., Gheerbrant, A. (1999) *Dicionário de Símbolos – Mitos, Sonhos, Costumes, Gestos, Formas, Figuras, Cores, Números.* Rio de Janeiro: José Olympio Editora.

Eliade, M. (2002) *O Xamanismo e as Técnicas Arcaicas do Êxtase.* São Paulo: Martins Fontes.

Hancock, G. (2007) *Supernatural: meetings with the ancient teachers of mankind.* New York: Disinformation Company Ltd.

Harner, M. (1982) *The Way of the Shaman: a guide to power and healing.* New York: Bantam Books.

Horowitz, M.J. (1975) *Hallucinations: An information-processing approach. In: Hallucinations. Behavior, experience, and theory.* Edited by Siegel, R.K., West, L.J. New York: John Wiley & Sons.

Huxley, A. (2004) *As Portas da Percepção – Céu e Inferno.* São Paulo: Globo S.A.

Kluver, H. (1966) *Mescal and Mechanisms of Hallucination.* Chicago: University of Chicago.

Lewis-Williams, D. (2004) *The Mind in the Cave: Consciousness and the Origins of Art.* London: Thames & Hudson.

Lewis-Williams, D. & Pearce, D. (2005) *Inside the Neolithic Mind: Consciousness, Cosmos and the Realm of the Gods.* London: Thames & Hudson.

Mikosz J.E. (2014) *Arte Visionária – Representações Visuais Inspiradas nos Estados Não Ordinários de Consciência (ENOC).* Curitiba: Editora Prismas.

Reichel-Dolmatoff, G. (1978) *O Contexto cultural de um alucinógeno aborígene: Banisteriopsis caapi.* [Online] Repositório do ISPA. Available from: http://hdl.handle.net/10400.12/1944. [Accessed: 21ᵗʰ June 2014].

DMT RESEARCH FROM 1956 TO THE EDGE OF TIME

ANDREW R. GALLIMORE AND DAVID P. LUKE

From a representative sample of a suitably psychedelic crowd, you'd be hard pressed to find someone who couldn't tell you all about Albert Hofmann's enchanted bicycle ride after swallowing what turned out to be a massive dose of LSD. The world's first acid trip (Hofmann, 1980) has since become a cherished piece of psychedelic folklore. Far fewer, however, could tell you much about the world's first DMT (N,N-dimethyltryptamine) trip. Although less memorable than Hofmann's story, it was no less important. The folklore would come later and reveal itself to be far weirder than anyone could have predicted. A DMT trip is certainly one of the most bizarre experiences a human can undergo and, although six decades have passed since the very first DMT trip, the experience continues to confound and remains fertile ground for speculation regarding its significance and meaning (Meyer, 1997; Luke, 2011; Gallimore, 2013). Of course, it would be extremely Western-centric to ignore the use of DMT by indigenous Amazonians in the ayahuasca brew (Shanon, 2003; Frenopoulo, 2005; Shanon, 2005; Schmidt, 2012) or the *cohoba* snuff (Schultes, 1984), but it was only after the effects of the pure compound were discovered that its role in these traditional preparations became clear.

There has been a resurgence of interest in DMT in the last couple of decades, largely inspired by the baroque orations of the late psychedelic bard Terence McKenna, who regarded DMT as "the secret"; producing the most intense and bizarre experience a human could have "this side of the yawning grave". Furthermore, although the endogenous production of DMT in humans has been established for several decades (Barker et al., 2012), attracting speculation as to its role in humans (Callaway, 1988; Wallach, 2009; Gallimore, 2013), recent research has provided more definitive evidence for a true functional role in human physiology (Frecska et al., 2013;Szabo et al., 2014). Dr Rick Strassman's groundbreaking study of the effects of DMT in humans (Strassman et al., 1994; Strassman, 1995; Strassman et al., 1996) has been a particularly potent catalyst for speculation regarding the significance of this unique psychedelic. Strassman, a psychiatrist at the University of New Mexico's School of Medicine, recruited 60 volunteers, the majority of whom received more than one dose of DMT by intravenous injection. This study was particularly special in that the subjective experiences of the volunteers took centre stage, with every detail of their trip narratives carefully recorded and many subsequently featuring in Strassman's psychedelic classic, *DMT: The Spirit Molecule* (Strassman, 2001). Whilst being the most ambitious and extensive study of DMT in humans, it certainly wasn't the first. For that, we'll need to go back a few decades.

THE FIRST DMT TRIP

The story begins sometime in 1953. Hungarian physician and chemist, Dr Stephen Szára, was planning a study to investigate possible biochemical factors in the aetiology of schizophrenia (Szára, 1989). News of the remarkable mind-bending effects of Hofmann's lysergic acid derivative had already spread throughout the European medical community and Szára was keen to procure a small supply to use in his own research. He wrote to Sandoz, the only source of LSD at the time, to place an order. However, Hungary was firmly locked behind the Iron Curtain and Sandoz seemed wary of sending him the potent new drug. His request was politely refused. Szára needed an alternative:

"At this point, I sent an order to a British pharmaceutical house to purchase 10 grams of mescaline. To my surprise, and delight, the drug arrived in December 1955. I remember weighing out 400 mg of mescaline in the laboratory a few days before Christmas and took it home..."
(Szára, 2014)

Having read and admired Huxley's *The Doors of Perception* (Huxley, 1954), Szára was keen to experience the effects of the drug himself. His choice of timing for his first mescaline trip now seems beautifully apposite:

"On Christmas Day I took it about 3pm. After about one hour I felt nothing, so I decided to go to the church on the top of the Castle-Hill in Budapest. On my way, on the bus, I started to feel that my vision had started to change; I was looking out through the window to the familiar landscape and seeing the trees moving in a strange way. When I got to the church I managed to get in, already full with people, standing room only. The ceremony had already started, loud organ music filled the air... to my surprise, as I was looking down to the marble floor, it was enlarging around me into a large circle, my neighbours seemingly far away, while I knew that I could touch them..."
(Szára, 2014)

Suitably heartened by the experience, Szára turned his attention to a recently published article by a trio of analytical chemists, Fish, Johnson and Horning (1955), on the chemical constituents of the *cohoba* snuff, used by indigenous South American tribes to induce states of religious Ecstasy (Schultes, 1984). Their analysis yielded only two major components; the first of these was the well-known toad skin secretion, bufotenine (Chilton et al., 1979). In typical 1950s style, medical double-act Fabing and Hawkins (1956) had already established the distinctly unpleasant, somewhat toxic and unimpressively psychoactive effects of this particular tryptamine by squirting large doses into a selection of unfortunate Ohio State Penitentiary inmates. Based on this data alone, it was generally assumed that bufotenine was responsible for the psychoactive effects of *cohoba*. Szára, however, was unconvinced.

The other major component of the snuff was the closely-related and pharmacologically unexplored alkaloid, N,N-dimethyltryptamine (DMT). It would be nice to write that Szára had a *peculiar presentiment* that DMT was the active psychedelic component of *cohoba*, but he didn't. However, it was obvious that those seeking communication with the gods were hardly likely to be impressed by the dangerously hypertensive, choking and nauseating effects of bufotenine. DMT was the only alternative and so Szára decided to make some. As well as being a physician, Szára has a Ph.D in organic chemistry and, using the recently published synthesis by Speeter and Anthony (1954), was able to synthesise ten grams of DMT within a few days. Unlike his American counterparts, Szára chose himself as the first test subject (actually, he chose a cat, but we'll skip that part). Mindful that Hofmann had considered 250μg of LSD a conservative first dose and ended up, in Szára's words, "bombed out", he opted for the same tiny amount, which he ingested orally. Of course, nothing happened. Over the next few days he gradually increased the dose up to 10mg/kg, or about three quarters of a gram. Still, no effect. Szára was somewhat discouraged and perhaps ready to abandon DMT when a colleague suggested that he should try injecting it:

> *"In April of 1956 (the exact day is unknown* [Author note: April has a special significance in psychedelic history, with Hofmann first taking LSD on the 19ᵗʰ of April]*), I tested three doses intramuscularly, paced at least two days apart to allow the drug to clear my body. The first dose (30 mg, around 0.4 mg/kg) elicited some mild symptoms – dilation of the pupils and some coloured geometric forms with closed eyes were already recognizable. Encouraged by these results, I decided to take a larger dose (75 mg, around 1.0 mg/kg), also intramuscularly. Within three minutes the symptoms started, both the autonomic (tingling, trembling, slight nausea, increased blood pressure and pulse rate) and the perceptual symptoms, such as brilliantly coloured oriental motifs and, later, wonderful scenes altering very rapidly."* (Szára, 2014)

Although Szára is unable to recall more details of this first experience, the trip had the typical DMT flavour familiar to contemporary users – complex geometric patterns give way to fully formed, immersive hallucinations. It was clear to Szára that this was the secret:

> *"I remember feeling intense euphoria at the higher dose levels that I attributed to the excitement of the realization that I, indeed, had discovered a new hallucinogen..."*　　　　　*(Szára, 2014)*

STRANGE NEW WORLDS: SZÁRA'S FIRST STUDY

Szára wasted no time in beginning the very first study of the effects of DMT in human subjects. He recruited 30 volunteers, mainly doctors from the hospital where he worked, the National Institute for Mental and Nervous Diseases, Budapest. All received 0.7 mg/kg (about 50mg for an average person) DMT intramuscularly and their experiences were carefully recorded (Sai-Halasz et al., 1958). Sadly, however, only a handful of these early DMT trip reports were published, purely as representative examples. The hospital has long since closed down, Szára's contemporaries have all since passed away and all the original data has been lost. Despite this, the few surviving reports offer an invaluable insight into the experiences of those very first volunteers. For anyone familiar with modern DMT trip reports or, indeed, who has taken DMT themselves, these early accounts will resonate. However, rather than simply presenting these reports in isolation, it might be more interesting to contextualise them with selections from modern trip reports for comparison. It must be borne in mind that although, to modern DMT users, the phenomenological content is certainly the most fascinating feature of the experience, the early DMT researchers were less aware of its significance. Szára, a self-confessed "old-fashioned scientist" and somewhat wary of "New Age stuff", pointed out:

> *"When these experiences, such as God, strange creatures and other worldliness, appeared in our DMT studies, we did not philosophize about them but, as psychiatrists, we simply classified them as hallucinations."*　　　　　*(Szára, 2014)*

As such, it is unsurprising that subjects weren't pressed to elaborate on their visions and the reports can seem terse. In contrast, trip reports garnered from online databases and forums, such as Erowid or the DMT Nexus, are often characteristically detailed and sometimes frankly verbose. Despite this difference, the parallels between these early reports and those of modern users are nonetheless revealing. In Szára's first study (Sai-Halasz et al., 1958), a 28-year-old male physician (we'll call him Adam) was one of the first to receive his 50mg dose:

> *"The room is full of spirits... the images come in such profusion that I hardly know where I want to begin with them! I see an orgy of color, but in several layers one after the other... Everything is so comical... one sees curious objects, but nevertheless everything is quickly gone, as if on a roller-coaster."* (Sai-Halasz et al., 1958)

Even this brief trip report extract features a number of fairly characteristic DMT motifs – the user is hurtled through a rapidly changing procession of complex visual imagery, meets discarnate entities (it is perhaps a reflection of the time that these were reported as 'spirits') and sees "curious objects". The rollercoaster analogy is often used to describe the sense of moving extremely rapidly through the highly complex visuals as the trip unfolds. The following extract is from one of 340 modern DMT trip reports collected by Peter Meyer and an anonymous blogger known only as Pup, compiled from numerous sources and published online (Meyer and Pup, 2005):

> *"This was by far the most intense experiment that I had done and it was like riding a roller coaster through a fractal. As the trip was winding down I tried to concentrate on the designs as they flowed by and through me to check out the complexities. As one of the more interesting designs flowed by I focused on a circular design that morphed as I focused on it into an eye with a grinning mouth below it. The smile seemed more maniacal than friendly, but was never less an amazing sight."*

Many modern users find that, whilst being indescribably bizarre, the experience is often suffused with a sort of comical ambience – the maniacal grinning mouth in the above report is perhaps indicative of this – and it's notable that Szára's subject, Adam, also seemed to experience this. Timothy Leary, in his seminal 1966 article, 'Programmed Communication During Experiences with DMT' (Leary, 1966), articulated this type of experience with characteristic eloquence:

> *"Eyes closed... suddenly, as if someone touched a button, the static darkness of retina is illuminated... enormous toy-jewel-clock factory, Santa Claus workshop... not impersonal or engineered, but jolly, comic, light-hearted. The evolutionary dance, humming with energy, billions of variegated forms spinning, clicking through their appointed rounds in the smooth ballet..."* *(Leary, 1966)*

This contributor to the fabulous drug information website, Erowid, describes a strikingly similar scene:

> *"It was generally like a wacky toy factory. Gadgets, widgets, twirling machines, stair-step pattern, Escher-like "space" and tunnels and chutes. The beings would seem to go "look!" and I felt I was supposed to look."* *(Erowid Experience #11258, 2001)*

Even in the hospital setting of Strassman's study, such madcap scenes were commonplace:

> *"Something took my hand and yanked me. It seemed to say, 'Let's go!' Then I started flying through an intense circus-like environment... there was a crazy circus sideshow – just extravagant. It's hard to describe. They looked like Jokers. They were almost performing for me. They were funny looking, bells on their hats, big noses."* *(Strassman, 2001)*

It is tempting to suggest that the "curious objects" reported by Adam are those that appear throughout the modern DMT trip report literature, often presented or revealed by elfish entities (we'll come to these later!):

> *"A gaggle of elf-like creatures in standard issue Irish elf costumes,*

complete with hats, looking like they had stepped out of a hallmark cards 'Happy Saint Patrick's Day' display, were doing strange things with strange objects that seemed to be a weird hybrid between crystals and machines." *(Meyer and Pup, 2005)*

"I saw the most indescribably beautiful 'objects' here. I was fascinated. The 'elves' seemed to want me to do the same thing that they were doing. It was frustrating." *(Meyer and Pup, 2005)*

EARLY ENTITY EXPERIENCES

Of course, to anyone at all familiar with the DMT experience, it is not so much *what* you see that commands particular attention but, rather, *where* you go and *who* you meet there. A large proportion of DMT users, about 50% in Strassman's study (Strassman, 2008), report travelling to normally invisible worlds and meeting an array of peculiar beings. Curiously, this type of experience was represented in Szára's first cohort (Sai-Halasz et al., 1958), so might be considered as a core feature of DMT experiences, rather than a later counter-cultural affectation. A 27-year-old female physician (we'll call her Bella) describes the characteristic auditory effects that precede 'breaking through':

"The whistling has stopped; I have arrived. In front of me are two quiet, sunlit Gods. They gaze at me and nod in a friendly manner. I think they are welcoming me into this new world."

(Sai-Halasz et al., 1958)

Although this new world seems somewhat more sedate than the worlds many DMT users are familiar with, the sense of having arrived in another place is unambiguous, as is the presence of non-human entities. There is even an attempt at communication:

"One of the Gods – only his eyes are alive – speaks to me: 'Do you feel better?'" *(Sai-Halasz et al., 1958)*

It's intriguing that Bella described these entities as "Gods", and tempting to speculate that they possessed qualities that inspired a sense of supreme power and wisdom. Entities with such qualities, whilst only occasionally referred to as gods, are certainly common in the experiences of modern DMT trippers:

> *"There were these beings that seemed to inhabit this place, that seemed to come off as vastly more intelligent and vastly more capable."*
>
> *(Erowid Experience #52797, 2006)*

Sometimes these entities actually claim to have the sort of power we might expect from Gods – the power to create life – although their behaviour can often be less than godly:

> *"I did see intelligent insect alien god beings who explained that they had created us, and were us in the future, but that this was all taking place outside of linear time. Then they telepathically scanned me, fucked me, and ate me."*
>
> *(Meyer and Pup, 2005)*

Later in the experience, Szára's subject, Bella, moved into an enclosed environment:

> *"From the darkness I see through the black iron lattice into the bright temple."*
>
> *(Sai-Halasz et al., 1958)*

Whilst no further details are given regarding this temple, many modern DMT users report entering a place with a temple-like quality:

> *"There is a corridor with a very tangible ambience, one can feel the space around. It now appears to be a temple structure of some futuristic sort, like some space age Hindu/Mayan temple with the walls displaying architecture similar to the Pyramid of the Sun at Teotihuacan except the walls are inverted to angle outward with the terraces reversed. It seems very real but also very fleeting, changing rapidly."*
>
> *(Meyer and Pup, 2005)*

Despite being somewhat brief and lacking detail, these earliest trip reports seem to hint at the type of experiences DMT users would describe decades later. This pattern continues with the studies that followed.

HORRIBLE ORANGE PEOPLE

Once Szára's first study was published, other physicians understandably became interested in DMT and began similar studies of their own. Turner and Merlis (1959) were interested in comparing the effect of pure DMT with those of the *cohoba* snuff. Their plucky recruits were convinced to inhale up to a gram of the snuff every 30 minutes, presumably until they could take no more. Unfortunately, neither subjective nor objective effects were observed, barring an awful lot of coughing and sneezing and general "discharging". DMT injections were more successful, although only a single trip report apparently merited publication. A 33-year-old psychotic female (we'll call her Carla) was given 50mg DMT intramuscularly. Following a brief period of anxiety and an apparent struggle against the effects, Carla fell into a dreamlike state, before awakening suddenly after about five minutes. She appeared to regain awareness of her surroundings and was then able to relate her experience to the attending physician:

> *"It is as if I were away from here for such a long time... In a big place and they were hurting me. They were not human. They were horrible... I was living in a world of orange people..."*
>
> *(Turner and Merlis, 1959)*

Of all the early DMT trip reports, this is one of the most intriguing. It isn't unusual for DMT users to experience entities that appear less than benevolent, although it should be pointed out that Carla had a history of physical abuse. A number of Strassman's volunteers also reported non-human entities with some degree of malevolent intent or, at least, that were either visually objectionable or performing some unpleasant act on the user. Insectoid creatures appear regularly in the literature:

> *"When I was first going under there were these insect creatures all around me. They were clearly trying to break through. I was fighting letting go of who I am or was. The more I fought, the more demonic they became, probing into my psyche and being. I finally started letting go of parts of myself, as I could no longer keep so much of me together."*
>
> *(Meyer and Pup, 2005)*

"With its innumerable eyes it gazed at me steadily and extended a tentacle. At the same moment it fired a beam of light directly between and above my eyes. The alien laser was pinkish-green. It hurt. I begged it to stop. I whimpered. Please stop, you're hurting me..."

(Meyer and Pup, 2005)

Sometimes, the user appears to be the subject of some sort of experimentation:

"There were four distinct beings looking down on me, like I was on an operating-room table. I opened my eyes to see if it was you and Josette, but it wasn't. They had done something and were observing the results. They are vastly advanced scientifically and technologically."

(Strassman, 2001)

"I felt like I was in an alien laboratory, in a hospital bed like this, but it was over there. A sort of landing bay, or recovery area. There were beings. I was trying to get a handle on what was going on. I was being carted around. It didn't look alien, but their sense of purpose was. It was a three-dimensional space... They had a space ready for me. They weren't as surprised as I was. It was incredibly un-psychedelic. I was able to pay attention to detail. There was one main creature, and he seemed to be behind it all, overseeing everything. The others were orderlies, or dis-orderlies."

(Strassman, 2001)

It was these 'alien experimentation' experiences that prompted Strassman to suggest, as others have, a relationship between DMT and alien abduction, during which the abductee is often subjected to painful experimental procedures, probing and measurements (Mack, 1994). However, despite some similarities, there are enough differences, including the absence of the classic 'alien greys' in the DMT state, to suggest that they are distinct (Luke, 2011). Although we have no further details as to the nature of the creatures that were hurting Turner & Merlis' subject, Carla, it's notable that accounts of negative interactions with non-human entities are not purely a modern feature of the DMT experience.

ENTER THE LITTLE PEOPLE: SZÁRA'S SECOND STUDY

As well as 'normal' subjects, Szára was also keen to observe the effects of DMT in some of the psychiatric patients at the hospital. There was a growing belief in psychiatry that these new "psychotogenic" agents, specifically LSD and mescaline, might be useful in the treatment of psychosis or at least as diagnostic tools. Szára selected 24 female in-patients, the majority with a diagnosis of schizophrenia, and all were given 1mg/kg DMT intramuscularly (Boszormenyi and Szára, 1958). Only three case reports were featured in the resulting publication and these were objective accounts of the patients' behaviour after DMT administration, as recorded by the physician – any insight into the subjective experiences of the patients could only be gleaned from their spontaneous utterances. Despite this, one of the case reports is particularly salient. The patient was a 30-year-old female (we'll call her Daisy) with "persecutory delusion and paranoid behaviour". A few minutes after the DMT was administered, the typical auditory effects began:

"She complains of a strange feeling, tinnitus, buzzing in the ear..."

(Boszormenyi and Szára, 1958)

It's common for a buzzing or humming sound to accompany the initial stages of a DMT trip, an effect also observed during out-of-body experiences (Blackmore, 1989). With Daisy, a period of agitation and confusion followed – "she keeps asking, 'Why do I feel so strange?'" – and then, after about 30 minutes, she seemed to indicate some loss of body awareness, also common in DMT users, which is echoed later in the session:

"As if my heart would not beat, as if I had no body, no nothing..."

(Boszormenyi and Szára, 1958)

Then, around 38 minutes into the session, her head begins to clear and she is able to recount her visions:

"I saw such strange dreams, but at the beginning only... I saw strange creatures, dwarfs or something, they were black and moved about..."

(Boszormenyi and Szára, 1958)

Reports of 'little people', described variously as elves, dwarfs, sprites or similar, are not only common during a DMT trip (Luke, 2013), they are perhaps one of its defining features. Of course, not everyone meets such beings, but 'the elves' are certainly the most famous denizens of the DMT realm. Terence McKenna's expositions on these highly animated little creatures, which he dubbed 'machine elves', are legendary:

> *"Trying to describe them isn't easy. On one level I call them self-transforming machine elves; half machine, half elf. They are also like self-dribbling jewelled basketballs, about half that volume, and they move very quickly and change. And they are, somehow, awaiting. When you burst into this space, there's a cheer!"*   *(McKenna, 1993)*

McKenna also called them 'tykes', which perfectly captures their spritely and mischievous nature. Whilst ubiquitous, they appear in a variety of forms, ranging from amorphous light beings to the classic elves of Germanic or Celtic folklore. Despite this variability, they seem to be unified by their character:

> *"The new geometry began to unfold layer after layer of laughing, giggling, incredibly lively beings... greeting me with enthusiastic cheers... the countless wonderful, hilarious, animated self-transforming liquid light energy creatures vied for my attention... They actually all start waving and saying 'goodbye' and 'Time to go, nice seeing you, Love you...'"*   *(Erowid Experience #85120, 2010)*

They even made an appearance in Strassman's study:

> *"That was real strange. There were a lot of elves. They were prankish, ornery, maybe four of them appeared at the side of a stretch of interstate highway I travel regularly. They commanded the scene, it was their terrain! They were about my height. They held up placards, showing me these incredibly beautiful, complex, swirling geometric scenes in them... They wanted me to look! I heard a giggling sound – the elves laughing or talking at high-speed volume, chattering, twittering."*   *(Strassman, 2001)*

Cott and Rock (2008) used thematic analysis to delineate the common themes within the DMT experience using anonymous volunteers. Although the accounts of only 19 respondents were analysed, the elves feature in one description of a scene that seems straight from a Brothers Grimm tale:

> *"Once I entered a room to see what looked like little elves working hard... I was watching these little guys work very hard on a bench, and they were building something."*   *(Cott and Rock, 2008)*

The elves, it seems, are everywhere...

ARCHETYPES OF THE S/ELF

Owing to Terence McKenna's almost Godlike eminence within the psychedelic community, and the gleeful enthusiasm with which he recounts their bizarre antics, it has become rather straightforward to dismiss the ubiquity of the elves as resulting from a sort of 'McKenna effect' – the little tykes ingress the subconscious of all who hear him speak, only to burst wildly into the frame shortly after the third toke. However, the appearance of animated dwarf-like creatures in at least one of Szára's subjects, when Terence McKenna was but a sprite himself, might hint that such beings represent something more universal, and that perhaps McKenna wasn't too far from the truth when he later claimed that "everybody gets elves". Whatever their true nature, the elves have understandably attracted a range of interpretations, from simple hallucinations to truly autonomous sentient beings from an alternate dimension. Somewhere in between, the elves are sometimes explained as latent Jungian archetypes scuttling from deep within the collective unconscious (Luke, 2011). This is the explanation that Szára favours:

> *"To explain the possible origins of dwarf or elf-like creatures, I wouldn't look for parallel universes or the "Quantum Sea" for explanation, but right in the brain, deeper than conscious memories..."*

(Szára, 2014)

Jung explained archetypes as "ever repeated typical experiences... stamped on the brain for aeons" (Jung, 1953). Anthropologist Charles Laughlin argues that these archetypes are encoded in neural structures present from birth and responsible for the experience of the foetus and infant (Laughlin, 1996). According to Laughlin, these "neurognostic structures" are both inheritable and subject to evolution (Laughlin, 2000). Drawing on this theory, Gallimore (2013) proposed that DMT might be an ancestral neuromodulator that was once secreted by the brain in psychedelic concentrations during sleep and allowing access to, or the development of, neurognostic structures (encoded as patterns of neural connectivity) entirely separate from those generating the experience of the consensus world (this one!). Whilst this function is apparently now lost, smoking or injecting DMT may allow these ancient neurognostic structures to be 'reactivated', allowing us access to a world that is not so much alien, but from which we have become alienated. Szára has a comparable idea, suggesting that certain ancient archetypal structures might be suppressed during normal consciousness, with DMT allowing them to break through to the surface:

> *"C.G. Jung's archetypes and symbols come to my mind as possible images, stored in neuronal connectivity patterns early in development. What DMT might do in adults is to slow down and stop reality testing (via the fronto-parietal loop) and let the Default Mode Network release the stored images and symbols into the perceptual system. It is the brain that stores and releases archetypal images into our altered consciousness..."* (Szára, 2014)

He proposes that the elves might represent one of the most fundamental archetypes:

> *"I would suggest that the dwarfs and elves, that appear in many, if not most of the DMT experiences, are in fact symbols of one of the most significant archetypes: 'the Self', in the form of circles or Mandalas released from the Collective Unconscious. They may be*

projections of early 'Selves' [Author note: we believe the pun to be intended] *stored in the infancy of the individual when ancestral DMT dominated brain functions..."* *(Szára, 2014)*

Of course, the idea that DMT somehow awakens deeply embedded archetypal structures remains highly speculative. However, Szára suggests that indirect evidence might already exist:

"There are actually some studies in neonatal rats (Beaton and Morris, 1984), showing that DMT can be detected at birth in low level, increasing significantly by day 12, staying high until day 31 and decreasing after this for the rest of their life. This is in line [with the model] postulated by Gallimore for the ancestral role for DMT. Obviously, this kind of data should be replicated in rats and in other animals as well as in humans if possible before the hypothesis gains credibility..." *(Szára, 2014)*

If the brains of human neonates secrete significant quantities of DMT, albeit temporarily, is it possible that the newborn child is able to access the DMT realm, but loses this ability long before memories can be laid down? It is intriguing that many DMT users feel a profound sense of déjà vu upon breaking through:

"The DMT space has a familiar feel to it. When I go to the DMT space, I often think, now I remember, this is where I have been before..." *(Meyer and Pup, 2005)*

The experience of déjà vu in healthy individuals is thought to result from a disruption of the normal *familiarity signal* that underlies recognition (O'Connor and Moulin, 2013). However, there is no agreement amongst psychologists and cognitive neuroscientists as to whether there is always an overlap between the perceptual experience (the DMT experience in this case) and a previously stored representation, and thus whether the experience of déjà vu stems, at least partly, from an actual memory (O'Connor and Moulin, 2010). Perhaps, in the case of DMT, it's not so much déjà vu as access to an authentic, albeit latent, memory trace. Perhaps we really *have* been there before. Maybe this is why the elves often welcome the tripper back with great celebratory uproar:

> *"They kept saying welcome back and words like: the big winner, he has returned, welcome to the end and the beginning, you are The One! As I looked around the room I felt the sense of some huge celebration upon my entry to this place. Bells were ringing, lights flashing..."*
>
> *(Erowid Experience #1839, 2000)*

It's rather uncanny that many of the themes that regularly appear in the DMT state are those we might naturally associate with childhood – clowns, jesters and jolly elves, playrooms and nurseries, fairgrounds and strange mechanical toys. Are small preschool children naturally attracted to small lively giggling characters (the outrageously popular Teletubbies, for example) because they remind them of similar creatures from the DMT realm? Something to think about.

IS THERE SOMETHING-IT'S-LIKE-TO-BE A MACHINE ELF?

Explaining the elves and other DMT entities as latent Jungian archetypes or other structures from deep within the collective unconscious doesn't seem to bring us any closer to knowing whether or not they can be considered *real*. Perhaps we need to look deeper. Jung himself proposed that fragments of the psyche, buried in the unconscious, might carry on a completely separate existence from the main complex. These *autonomous psychic complexes* form a miniature, self-contained psyche and are, perhaps, *even capable of a consciousness of their own* (Jacobi, 1959). This is a powerful idea. Descartes famously failed to deny his own existence, or at least the existence of his own mind. Neuroscientist Giulio Tononi seems to generalise this existential undeniability to anything possessing subjective consciousness. According to Tononi, whilst many things might be considered real, if something exists from its own subjective perspective – if there is *something-it's-like-to-be it* – then it's *really real* (Tononi, 2014); its reality cannot be denied. So, with regards to the reality or otherwise of the elves, perhaps we should ask: *is there something-it's-like-to-be a machine elf?* If Jung's autonomous complexes are to be taken seriously, then the answer is a very big and startling maybe. Maybe! And, if so, then the elves might not just be real, but *really real!* This might suggest that the elves could have intentionality and are actively seeking to communicate with us, even if the communicator turns out to be an alienated fragment of the communicatee. As McKenna (1991) suggests, "we are alienated, so alienated that the self must disguise itself as an extraterrestrial in order not to alarm us with the truly bizarre dimensions that it encompasses. When we can love the alien, then we will have begun to heal the psychic discontinuity that [plagues] us."

In an apparent attempt to transcend such philosophical speculation, computer scientist Marko Rodriguez proposed a methodology for experimentally testing whether or not DMT entities objectively exist – giving them advanced mathematical problems to solve (specifically, finding the unique prime factors of very large numbers) (Rodriguez, 2007). Rodriguez assumes that the highly advanced DMT entities (they very often appear extremely intelligent and technologically advanced) would waste no time showing off their computational prowess by feeding the correct answers to the expectant tripper, thus proving their objective existence. However, as Luke (2011) points out, it can't be assumed that the tripper couldn't receive the correct answers from some earthly incarnate source by a so-called 'super-psi' effect. Despite this shortcoming, this type of experiment might eventually prove useful in extracting information from the DMT entities that one might struggle to ascribe to our earthbound domain. In fact, perhaps the extraction of useful information would be a better standard by which to judge the objective existence or otherwise of the DMT reality and its inhabitants. Terence McKenna often sidestepped questions regarding the objective existence or reality of the entities and simply asked them: *What can you show me?*

THINKING THE UNTHINKABLE

So far, we have avoided discussing the possibility that DMT might actually open a doorway to an alternate world and, for those of you rolling your eyes now, such an idea might be seen as frankly ludicrous. However, it can't be ignored that a large proportion of DMT users, from Szára's earliest subjects to modern amateur psychonauts, often arrive in the same type of world – *highly artificial, constructed, inorganic, and in essence technological* (Hancock, 2005) – and meet the same types of entities. If (albeit a big if) an experiment could be designed to extricate information from the DMT realm and 'super-psi' effects or other earthly sources could be ruled out, this might allow us to differentiate latent archetypes or other unconscious structures from a truly autonomous alternate reality. The results of such an experiment, if positive, would have profound implications for our understanding of reality. Obviously, it would be wildly astounding to discover that such a strange world exists and could well have existed long before our universe popped from nowhere (ahem) 14 billion years ago ("Give us one free miracle and we'll explain the rest..." as Terence McKenna used to quip). However, there's nothing in the laws of physics to rule out parallel worlds as such. The most astonishing revelation would not be the existence of such a world, but the fact that we had the ability to access it with such facility; by inhaling a couple of lungfuls of one of the simplest and most common alkaloids in the plant kingdom.

This revelation would force a far more fundamental paradigm shift in our understanding of reality and our place in it. The major problem with the *alternate world* explanation for DMT is what might be called the *data input problem.* There seems to be no obvious means for our brains to receive data from an alternate dimension and to explain such a phenomenon might require us to rethink the structure of reality itself. For example, the idea that we might live in a type of computer-simulated universe is actually now receiving serious academic consideration (Bostrom, 2003; Whitworth, 2007; Beane, Davoudi & Savage, 2012). Not only would many of the implicit quirks within the laws of physics be easily explained by a simulated universe (Whitworth, 2007), it might also be straightforward to explain how data normally disconnected from our reality program might be gated by a relatively simple subprogram (i.e. DMT). Perhaps, instead of looking for glitches or drifts in the physical constants of the universe (Barrow, 2003) to test the simulated universe theory, we ought to be looking for more explicit clues. And perhaps DMT is one of those clues. This would also naturally raise the question of intentionality – was the DMT subprogram deliberately embedded for us to find? And, if so, by whom and for what reason? We'll leave this question as an exercise for the reader.

Of course, for those with no personal experience with DMT, it's reassuringly straightforward to glibly dismiss alternate alien dimensions and disembodied intelligences as mere hallucination. Clever philosophical rhetoric, offbeat thought experiments and appeals to obscure Jungian manuscripts are unlikely to convince the DMT-naïve that there's something far more interesting going on. The only reliable convincer appears to be somewhere comfortable to lie down for ten minutes and a small glass pipe.

OH MY FUCKING GOD!

Beyond the overwhelmingly rich, complex and fascinating visual content of the DMT experience, the elves and their strange hyperdimensional habitat seem to percuss the user at a point deep in the core of their being. The immediate response is often a profound sense of shock and the returning tripper, eyes wide and shaking, might struggle to verbalise anything beyond repeating "OH MY FUCKING GOD!" What is particularly interesting about DMT is that this shock is caused, not just by the bizarre nature of the experience, but by an unshakeable feeling of authenticity – the individual is unable to deny the *reality* of the experience; unable to dismiss it as hallucination or repressed memories bubbling up from the darkest corners of the unconscious mind. The most fundamental ontological assumption – that this earthly reality is the one and only *real thing* – is instantaneously shattered with little hope of being restored. The DMT reality *feels* real and *is* real, even after returning to normal consciousness.

The late Harvard psychiatrist, Dr John E. Mack, conducted detailed interviews and hypnotic regression sessions with over 200 so-called alien abductees. Despite not suffering from any known neuropsychological pathology that he could identify (i.e. perfectly sane), all of the abductees were uncompromising in their insistence that their experiences were real; that they had really happened. Similar to returning DMT trippers, this feeling of absolute certainty collided with their most basic assumptions regarding what was and wasn't possible with such force that they were left in a state that Mack termed "ontological shock" (Mack, 1994). Humans, sane ones at least, are extremely good at distinguishing reality from fantasy and the vast majority are quick to accept that a particularly strange dream was just that – a dream and not real. Whilst this *reality-testing* is impaired during dreaming and psychosis (Limosani et al., 2011), waking from the dream or recovery from psychosis is generally sufficient to restore this important ego faculty and the dream or hallucination is recognised for what it is. This makes the DMT experience all the more compelling and paradoxical – it is far stranger than (almost) any dream and yet there remains a remarkable inability to shake the feeling that it was truly real once the experience has ended.

HARDER, DEEPER, LONGER

Despite almost six decades having passed since Szára's first study, we appear to be no closer to a definitive explanation for DMT's astonishing psychoactive effects. It's quite tempting to assume that the strangely characteristic DMT-like experiences that so many users report, from playful giggling elves to grotesque alien "pro-bono proctologists", can be explained in terms of modern cultural memes that propagate through underground psychedelic literature, archival Terence McKenna lectures and, most recently, through the internet. However, the few surviving trip reports from these early DMT studies perhaps suggest that things are a little more complicated than that and therein lay their importance. DMT evokes experiences with a highly characteristic flavour and content that appear to be somewhat independent of the cultural setting in which they manifest, coupled with a frighteningly compelling sense of authenticity. Whether this points towards a truly autonomous alternate universe to which DMT somehow gates access or towards some deeply embedded structures within the human collective unconscious (perhaps both), or something entirely different, these experiences are far from trivial to explain and are certainly worthy of proper academic study.

Although psychedelic drug research is now recovering following a several-decades-long hiatus, the academic focus now tends to be on the neural correlates of the experience, rather than the experience itself. Whilst this research is important, there is the danger that the psychedelic state will be unwittingly explained away, neatly packaged as an interesting variant of brain network activity. This is wholly unsatisfactory and it is important that objective neuroimaging data is complemented by detailed phenomenological studies of the DMT state. The authors are currently performing a detailed (and long overdue) quantitative phenomenological analysis of Strassman's original 'bedside notes' and such carefully recorded trip reports will likely form the primary data source in future research, although psychedelic researchers should be prepared to enter the DMT realm themselves with the aim of answering specific questions and performing experiments. After all, if you want a tiger's cub, you must go into the tiger's cave.

Terence McKenna exhorted psychedelic drug users to see themselves as explorers –

"You are an explorer, and you represent our species, and the greatest good you can do is to bring back a new idea, because our world is endangered by the absence of good ideas."

Whilst a handful of lay psychonauts might manage to dive deep enough to bring back a few good ideas, this is really the psychedelic equivalent of amateur free diving and is ultimately limited by the lung capacity and training of the individual and the brief duration of action of the drug. There is perhaps an attractive romanticism attached to the idea of sitting cross-legged on a brightly coloured, hand-woven rug, lighting incense and raising a hand-blown glass pipe to the lips. But DMT seems to demand more from us than this. It appears to carry a message so shockingly profound and important that it demands careful preparation and skilful use of our technological apparatus to successfully retrieve and decode it. DMT is unique amongst the classical psychedelics in not only being very short acting, but also not exhibiting a tolerance effect with repeated use (Strassman et al., 1996). As such, using the same technology developed for maintaining a stable brain concentration of anaesthetic drugs during surgery, it would be feasible to administer DMT by precisely regulated continuous intravenous infusion, permitting the explorer an extended, and theoretically indefinite, sojourn in the DMT reality (Gallimore, *manuscript in preparation*). Although a well-prepared strong ayahuasca brew might achieve a crudely comparable effect, standardisation of the dose is much more difficult and the concentration of DMT in the brain will fluctuate based on a range of pharmacokinetic and metabolic factors. A well-designed continuous IV protocol could account for these factors, allowing the brain DMT concentration to be kept stable or manipulated to gradually move the explorer deeper and deeper into the DMT space. This approach might give the brave voyagers enough time to orient themselves, get their intellective tools in order and cast their nets far out into the "dark ocean of mind" in the hope of bringing back the message that DMT has been trying to convey since Szára first dipped his toes in the water almost 60 years ago. If ever there was a time, it's now. Ahoy, shipmates! Ahoy!

Barker, S. A., Mcilhenny, E. H. & Strassman, R. (2012). A critical review of reports of endogenous psychedelic N, N-dimethyltryptamines in humans: 1955-2010. *Drug Testing and Analysis* 4, 617-635.

Barrow, J.D. (2003). Living in a simulated universe. Available at: www.simulation-argument.com/barrowsim.pdf.

Beane, S. R., Davoudi, Z., Savage, M.J. (2012). Constraints on the universe as a numerical simulation. arXiv:1210.1847v2.

Beaton, J. M. & Morris, P. E. (1984). Ontogeny of N,N-dimethyltryptamine and related indolealkylamine levels in neonatal rats. *Mechanisms of Ageing and Development* 25, 343-347.

Blackmore, S. J. (1989). *Beyond the Body: An Investigation of Out-of-Body Experiences*. Chicago: Academy Chicago Publishers.

Boszormenyi, Z. & Szára, S. (1958). Dimethyltryptamine experiments with psychotics. *Journal of Mental Science* 104, 445-453.

Bostrom, N. (2003). Are you living in a computer simulation? *Philosophical Quarterly* 53, 243-255.

Callaway, J. C. (1988). A proposed mechanism for the visions of dream sleep. *Medical Hypotheses* 26, 119-124.

Chilton, W. S., Bigwood, J. & Jensen, R. E. (1979). Psilocin, bufotenine and serotonin - historical and biosynthetic observations. *Journal of Psychedelic Drugs* 11, 61-69.

Cott, C. & Rock, A. (2008). Phenomenology of N,N-Dimethyltryptamine Use: A Thematic Analysis. *Journal of Scientific Exploration* 22, 359-370.

Fabing, H. D. & Hawkins, J. R. (1956). Intravenous bufotenine injection in the human being. *Science* 123, 886-887.

Fish, M., Johnson, N. M. & Horning, E. L. (1955). Piptadine alkaloids. Indole base of *P. peregrine* (L.) and related species. *Journal of the American Chemical Society* 77, 5892-5895.

Frecska, E., Szabo, A., Winkelman, M. J., Luna, L. E. & Mckenna, D. J. (2013). A possibly sigma-1 receptor mediated role of dimethyltryptamine in tissue protection, regeneration & immunity. *Journal of Neural Transmission* 120, 1295-1303.

Frenopoulo, C. (2005). The ritual use of ayahuasca. *Journal of Psychoactive Drugs* 37, 237-239.

Gallimore, A. R. (2013). Building Alien Worlds - The Neuropsychological and Evolutionary Implications of the Astonishing Psychoactive Effects of N,N-Dimethyltryptamine (DMT). *Journal of Scientific Exploration* 27, 455-503.

Hancock, G. (2005). *Supernatural: Meetings with the Ancient Teachers of Mankind*. Arrow Books, London.

Hofmann, A. (1980). *LSD: My Problem Child*. McGraw-Hill Book Company, New York.

Huxley, A. (1954). *The Doors of Perception*. Chatto & Windus, London.

Jacobi, J. (1959). *Complex/Archetype/Symbol in the Psychology of C. G. Jung*. Pantheon Books Inc, New York..

Jung, C. G. (1953). *Two Essays on Analytical Psychology*. Routledge & Kegan Paul, London.

Laughlin, C. D. (1996). Archetypes, neurognosis and the quantum sea. *Journal of Scientific Exploration* 10, 375-400.

Laughlin, C. D. (2000). Biogenetic structural theory and the neurophenomenology of consciousness. *Toward a Science of Consciousness Iii: the Third Tucson Discussions and Debates*, 459-473.

Leary, T. (1966). Programmed communication during experiences with DMT. *Psychedelic Review* 8, 83-95.

Limosani, I., D'agostino, A., Manzone, M. L. & Scarone, S. (2011). The dreaming brain/mind, consciousness and psychosis. *Consciousness and Cognition* 20, 987-992.

Luke, D. (2011). Discarnate entities and dimethyltryptamine (DMT): Psychopharmacology, phenomenology and ontology. *Journal of the Society for Psychical Research* 75, 26-42.

Luke, D. (2013). "So long as you've got your elf: Death, DMT and discarnate entities," in *Daimonic imagination: Uncanny intelligence*. Voss, A. & Rowlandson, W. (eds.) 1st ed. Cambridge Scholars Publishing, Cambridge. 282-291.

Mack, J. E. (1994). *Abduction - Human Encounters with Aliens*. Macmillan Publishing Company, New York.

McKenna, T. (1991) The Archaic Revival: Speculations on Psychedelic Mushrooms, the Amazon, Virtual Reality, UFOs, Evolution, Shamanism, the Rebirth of the Goddess & the End of History. Harper, San Francisco.

Mckenna, T. (1993). "Interview for OMNI magazine", in: *OMNI Magazine*.).

Meyer, P. (1997). *Apparent Communication with Discarnate Entities Induced by Dimethyltryptamine (DMT)* [Online]. http://www.serendipity.li/dmt/dmtart00.html. Available: http://www.serendipity.li/dmt/dmtart00.html [Accessed 20th August 2014].

Meyer, P. & Pup (2005). *340 DMT Trip Reports* [Online]. http://www.serendipity.li/340_dmt_trip_reports.htm. Accessed: 1 Oct 2014. Available: http://www.serendipity.li/340_dmt_trip_reports.htm [Accessed 1st October 2014].

O'connor, A. R. & Moulin, C. J. A. (2010). Recognition Without Identification, Erroneous Familiarity & Deja Vu. *Current Psychiatry Reports* 12, 165-173.

O'connor, A. R. & Moulin, C. J. A. (2013). Deja vu experiences in healthy subjects are unrelated to laboratory tests of recollection and familiarity for word stimuli. *Frontiers in Psychology* 4, 9.

Rodriguez, M. A. (2007). A methodology for studying various interpretations of the N,N-dimethyltryptamine-induced alternate reality. *Journal of Scientific Exploration* 21, 67-84.

Sai-Halasz, A., Brunecker, G. & Szara, S. (1958). Dimethyltryptamine: a new psycho-active drug (unpublished English translation). *Psychiatria et neurologia* 135, 285-301.

Schmidt, B. E. (2012). Ayahuasca, Ritual & Religion in Brazil. *Anthropos* 107, 276-277.

Schultes, R. E. (1984). 15 years of study of psychoactive snuffs of South-America - 1967-1982 - a review. *Journal of Ethnopharmacology* 11, 17-32.

Shanon, B. (2003). Altered states and the study of consciousness - The case of ayahuasca. *Journal of Mind and Behavior* 24, 125-153.

Shanon, B. (2005). *The Antipodes of the Mind - Charting the Phenomenology of the Ayahuasca Experience*. Oxford University Press, Oxford.

Speeter, M. E. & Anthony, W. C. (1954). The action of oxalyl chloride on indoles - a new approach to tryptamines. *Journal of the American Chemical Society* 76, 6208-6210.

Strassman, R. (2001). *DMT - The Spirit Molecule*. Park Street Press, Vermont.

Strassman, R. (2008). "The varieties of the DMT experience," *Inner Paths to Outer Space*. Strassman, R. (ed.) Park Street Press, Vermont., 51-80.

Strassman, R. J. (1995). Human psychopharmacology of N,N-dimethyltryptamine. *Behavioural Brain Research* 73, 121-124.

Strassman, R. J., Qualls, C.R. & Berg, L.M. (1996). Differential tolerance to biological and subjective effects of four closely spaced doses of N,N-dimethyltryptamine in humans. *Biological Psychiatry* 39, 784-795.

Strassman, R. J., Qualls, C. R., Uhlenhuth, E. H. & Kellner, R. (1994). Dose-response study of N,N-dimethyltryptamine in humans .2. Subjective effects and preliminary results of a new rating scale. *Archives of General Psychiatry* 51, 98-108.

Szabo, A., Kovacs, A., Frecska, E. & Rajnavolgyi, E. (2014). Psychedelic N,N-Dimethyltryptamine and 5-Methoxy-N,N-Dimethyltryptamine Modulate Innate and Adaptive Inflammatory Responses through the Sigma-1 Receptor of Human Monocyte-Derived Dendritic Cells. *Plos One* 9, 12.

Szára, S. (1989). The social chemistry of discovery - the DMT story. *Social Pharmacology* 3, 237-248.

Szára, S. (2014). *Interview conducted by email on the discovery of DMT and speculations regarding the phenomenology of the experience.* Gallimore, A.R. & Luke. D.P. , (eds.)

Tononi, G. (2014). *RE: personal communication.*

Turner, W. J. & Merlis, S. (1959). Effect of some indolealkylamines on man. *Archives of Neurology and Psychiatry* 81, 121-129.

Wallach, J. V. (2009). Endogenous hallucinogens as ligands of the trace amine receptors: A possible role in sensory perception. *Medical Hypotheses* 72, 91-94.

Whitworth, B. (2007). The physical world as virtual reality. *CDMTCS Research Report Series* 316, 1-17.

CONTRIBUTORS

ANDREW R. GALLIMORE, PH.D., is a neurobiologist, pharmacologist, and chemist currently based at the Okinawa Institute of Science and Technology. He has been interested in the neural basis of psychedelic drug action for many years and is the author of a number of articles on DMT and the psychedelic state, including Building Alien Worlds (2013), in which he developed a new model of DMT's effects on neural function and its relationship to human neuroevolution. He is currently collaborating with Dr David Luke and Dr Rick Strassman to perform the first detailed phenomenological analysis of the subjective reports of the 60 volunteers in Strassman's landmark human DMT study. It is hoped that this work will further our understanding of the ontological significance of the astonishing psychoactive effects of this unique psychedelic.

OLI GENN-BASH, M.A., has a Masters in Political Theory and Practices of Resistance from the University Of Kent, Canterbury. His research has included: the study of mysticism/mystical experience in relation to resistance against traditional religious practice, the study of philosophy and politics of truth, psychedelic thought as practice of resistance, and specific focus on the relationship between humour, stand-up comedy, and resistance. He is the former President of the University of Kent Psychedelics Society.

PETER SJÖSTEDT-H, M.A., is an Anglo-Scandinavian philosopher who specialises in the thought of Schopenhauer, Nietzsche, Bergson and Whitehead, and within the field of Philosophy of Mind. Peter has a Bachelor's degree in Philosophy and a Master's degree in Continental Philosophy from the University of Warwick. He became a Philosophy Lecturer in London for six years before recently returning to the tranquillity of westernmost Cornwall. He is now an independent philosopher – giving talks, publishing essays, and preparing to embark upon his Ph.D. www.philosopher.eu

DAVE KING, B.SC., is a founding Director of Breaking Convention and the founding President of the UKC Psychedelics Society. He organised the Shulgin Blotter Art Fundraiser in 2011-12, raising more than $20,000 for the Shulgins' medical bills. He has worked for the Beckley Foundation, the National University of Singapore, and the NHS, and has presented research at six international conferences.

SAM GANDY, MRES, is currently working on a Ph.D. in ecological sciences on termites, soils and ecosystem services, based between the James Hutton Institute in Aberdeen and southern Ethiopia. He helped establish the company beefayre and is a contributor to Reality Sandwich and PsyPress UK and in his free time enjoys writing and researching consciousness-related topics.

MARIA PAPASPYROU, M.SC., is a BACP accredited counsellor and psychotherapist. She has been working as a therapist for over ten years, in the fields of mental health and education, alongside her private practice. Entheogens and healing have been major reference points of interest for many years. In psychedelic science the two are able to join and she explores the sacramental and healing properties of entheogens, and how these can foster development and the growth of human and societal tacit potential.

BEN SESSA, M.B.B.S., is a psychiatrist specialising in substance misuse and post-trauma pathology in South West England. He is leading the UK's first clinical study using MDMA to tackle PTSD, in a Cardiff-based fMRI neuro-imaging study. He is the author of several books, including *The Psychedelic Renaissance*, and is a co-founder and chair of Breaking Convention. Ben is dedicated to exploring the evidence-based potential for psychedelic drugs to treat mental disorders and for individuals' personal growth and development.

MICHAEL MITHOEFER, M.D., is a psychiatrist in Charleston, SC. He and his wife, Annie Mithoefer, completed the first clinical trial of MDMA-assisted psychotherapy for treatment-resistant PTSD. They continue to conduce MDMA treatment research, including training and collaborating with

MDMA research teams in the US and in other countries. He is board certified in psychiatry, emergency medicine and internal medicine, is a Fellow of the American Psychiatric Association and Clinical Assistant Professor of Psychiatry at the Medical University of South Carolina.

RICK DOBLIN, PH.D., is the founder and executive director of the Multidisciplinary Association for Psychedelic Studies (MAPS). His doctorate dissertation, in public policy from Harvard's Kennedy School of Government, was on the regulation of the medical uses of psychedelics and his undergraduate thesis at New College of Florida was a 25-year follow-up to the classic Good Friday Experiment. His professional goal is to help develop legal contexts for the beneficial uses of psychedelics and marijuana and eventually to become a legally licensed psychedelic therapist.

HENRY DOSEDLA was engaged with continuous fieldwork as an archaeologist and social anthropologist during the early 1970s among the last societies representing Neolithic standards in Melanesia, dealing with their environment management, mythological folk biology, medical traditions and religious concepts including divination systems. He also was engaged in several development programmes and documented gradual stages of cultural change including related social effects. When in charge of the Prehistory Department of the German Museum of Agriculture at Hohenheim University, Stuttgart he was engaged in the development of local open air museums with reconstructions of traditional rural buildings. After retirement his further research and publications were focused on adequate parallels between recent archaic societies and conditions in prehistoric Europe.

JACK HUNTER, M.A., is a doctoral candidate in social anthropology at the University of Bristol. His research looks at contemporary British trance and physical mediumship, focussing on themes of altered consciousness, personhood and performance. He is the founder and editor of *Paranthropology: Journal of Anthropological Approaches to the Paranormal*, and co-editor with Dr David Luke of *Talking With the*

Spirits: Ethnographies from Between the Worlds. He is currently training to be a teacher of religious education with the University of Chester.

MATTHEW CLARK, PH.D., is a freelance academic, lecturer and researcher. He is currently a Research Associate affiliated to the School of Oriental and African Studies (University of London). He has published articles and books on Indian sadhus/sadhvis (holy men and women) and yoga. He is also a musician (Mahabongo).

WILLIAM ROWLANDSON, PH.D., is a Senior Lecturer in Hispanic Studies at the University of Kent, and former Director of the Centre for the Study of Myth. He has recently completed a book concerning Borges and mysticism, which examines the relationship between Borges' own recorded mystical experiences and his appraisal of Swedenborg and other mystics. William's work on Cuban poet and novelist José Lezama Lima concentrated on Lezama's equation of poetry and the numinous. With co-Director Angela Voss, William organised a conference at the University of Kent in May 2011 entitled Daimonic Imagination: Uncanny Intelligence.

CHIARA BALDINI is an independent researcher. Her work explores how altered states of consciousness have been embedded in different rituals over the course of Western history. She wrote 'Dionysus Returns: Contemporary Tuscan Trancers and Euripides' The Bacchae' featured in *The Local Scenes and Global Culture of Psytrance* (Routledge, 2010) and she co-authored with Graham St John 'Dancing at the Crossroads of Consciousness: Techno-Mysticism, Visionary Arts and Portugal's Boom Festival' for the Brill's *Handbook of New Religions and Cultural Production* (2012).

ANDY ROBERTS (meugher@gmail.com) is an historian of Britain's psychedelic culture, with an emphasis on LSD. He is the author of *Albion Dreaming: A popular history of LSD in Britain* and has contributed to numerous anthologies and journals. Andy has also written or co-written several books on the UFO mythos and British folklore. He is a feature writer and columnist for *Fortean Times*. Andy is currently working on a biography of Michael Hollingshead, the man who turned Timothy Leary onto LSD.

DEIRDRE RUANE, M.A. is a doctoral researcher at the University of Kent's School of Social Policy, Sociology and Social Research with a longtime interest in festival culture. Her academic interests include psychedelic harm reduction, intentional communities online and offline, and the anthropology of religion and ritual. Her Ph.D. research concerns peer-based psychedelic support and harm reduction projects at EDMC and other counter cultural festivals. She has volunteered as a sitter with these projects at events in the UK, US and Portugal.

GRAHAM ST JOHN, PH.D., is a cultural anthropologist and author of the forthcoming book *Mystery School in Hyperspace: A Cultural History of DMT* (North Atlantic Books, 2015), as well as several other books including *Global Tribe: Technology, Spirituality and Psytrance* (Equinox, 2012), *Technomad: Global Raving Countercultures* (Equinox, 2009), and *FreeNRG: Notes From the Edge of the Dance Floor* (CommonGround, 2001). He is Executive Editor of *Dancecult: Journal of Electronic Dance Music Culture*. His website is at www.edgecentral.net.

ALEXANDER BEINER, M.A. is a writer, psychedelic theorist and co-founder of Open Meditation, a London-based company that teaches non-religious mindfulness meditation. His first novel *Beyond the Basin* was well-received in the psychedelic community and an accompanying article to the novel was featured in *The Guardian* when the book came out. The article was one of the first in a national newspaper calling for drugs to be legalised in spiritual practice. Alexander is also the host of Visionary Artists Podcast, the world's most comprehensive archive of interviews with established and aspiring visionary artists.

ALLAN BADINER, M.A., is a student of Buddhism, a contributing editor at *Tricycle* magazine, and an ecological and psychedelic activist. He edited the books *Zig Zag Zen: Buddhism and Psychedelics* (Synergetic Press), Dharma Gaia: A Harvest in Buddhism and Ecology, and *Mindfulness in the Marketplace: Compassionate Responses to Consumerism* (Parallax Press). Allan holds a masters degree in Buddhist Studies from the College of Buddhist Studies in Los Angeles and serves on the boards of Rainforest Action Network, Threshold Foundation, and Project CBD.

JULIAN VAYNE is an occultist and the author of a number of books, essays and articles in both the academic and esoteric press. His name is most closely associated with chaos magick and he is also an initiated Wiccan and member of the Kaula Nath lineage. He has been involved with the magickal world for over 20 years. He has led a variety of esoteric workshops and courses and is a prominent figure in contemporary British occultism. His interests include drugs and magick, permaculture and the politics of sustainability, teaching and graphic art.

ROBIN MACKENZIE, PH.D., is the Director of Medical Law and Ethics at the University of Kent. She has published widely on psychoactive substance use, non-ordinary states of consciousness, end-of-life issues, decision-making capacities and neurodiversity.

JOSÉ ELIÉZER MIKOSZ, PH.D., was born in Curitiba, Brazil. He is a painter, an art instructor, and a visionary art researcher. His doctoral thesis explored Visionary Art and Ayahuasca, and was presented at the Interdisciplinary Program in Human Sciences. Mikosz's research deals with the interaction between art and consciousness, searching examples of the artists' expression of their visions achieved through non-ordinary states of consciousness. He is currently an art instructor and research supervisor at the State University of Paraná (Unespar-Embap). Group leader of Inter and Transdisciplinary Studies in Art, Consciousness and Related Poetics, and Editor of the International Interdisciplinary Journal of Visual Arts – *Art & Sensorium*. Member of the Scientific Committee of the Rose Croix International University (URCI-AMORC), member of the Advisory Board of the Research Centre for the Study of Psycointegrator Plants, Visionary Art and Consciousness – Wasiwaska, and Associated to the Center for Interdisciplinary Studies on Psychoactives – NEIP.

DAVID LUKE, PH.D. David is Senior Lecturer in Psychology at the University of Greenwich where he teaches the Psychology of Exceptional Human Experience. He was President of the Parapsychological Association between 2009-11 and has published more than 100 academic papers on the intersection of transpersonal experiences, anomalous phenomena and altered states of consciousness. He has co-authored/co-edited four books on psychedelics and paranormal experience, directs the Ecology, Cosmos and Consciousness salon at the Institute of Ecotechnics, and co-founded Breaking Convention.

ACKNOWLEDGEMENTS

DAVE KING Breaking Convention is an enormous undertaking that relies upon the support and effort of a large number of people, mostly volunteers. I would like to thank everyone who contributed to our second conference, from which these essays proceed. In particular, I would like to thank Anna Waldstein, a founding member of Breaking Convention who left the board of directors in late 2013 to focus her energies elsewhere. Without her huge contributions, the conference would not exist. Many thanks also to Mark Pilkington and the team at Strange Attractor Press for putting together such an elegant product. I am enormously grateful for the work and patience of every chapter author, especially those who had to put up with more than their fair share of my editorial suggestions.

Warm thanks also go to Graham Field, Bernard Flanagan and Fasil Hussain at Longmores Solicitors, without whom we may never have achieved charitable status.

DAVID LUKE My sincerest thanks to the innumerable people who helped to make Breaking Convention 2013 possible, not least my fellow Breakeros on the core team, Dave, Anna, Ben, Cam & Aimee, who make the whole crazy adventure of putting on such a massive event for no personal financial gain such a delirious adventure. Massive thanks also to the following, Blue Firth & Judith Way for the promotional art, all the other multitudinous artists and performers for presenting their work, especially Karen Barnes, Dave Bardo, Paul Friedlander, Maria & Nestor of Kimatica, Birah Rose (wow) and crew, Phantasmagoria and the oodles of speakers, musicians, workshop leaders and volunteers who gave generously of their time and expertise, Jane Colings for the cosmic cake. Extra massive thanks to Adam Malone & Sammy Fourway for organising London PsychFest, thanks too to the amazing psychedelic rock bands, especially Space Ritual for really blowing my mind. Uber thanks to Jonathan Greet for the amazing photography, and Giorgos

Mitropapas for designing the programme booklet, Cara Lavan & Hayley Cattlin and the volunteer film crew for the massive job of filming all the lectures, David Fuller for press liaison, and of course Mark Pilkington for publishing. Thanks too to Rodrigo Rurawe for bringing over the Huichol shaman Don Santos, and to Santos for his incredible ceremony. Ultimately my deepest thanks to my wife Anna Hope for all the moral (and immoral) support.

BEN SESSA Clear and obvious thanks goes out to Dave King for pulling together this fabulous volume of mind-munching essays. His job has been one of herding psychedelic cats – both the authors of the submitted manuscripts and the rest of us co-editors (Dave Luke, Cameron Adams, Aimee Tollan and myself), whose cognitive meanderings have carried our priorities in all kinds of obscure directions, whilst Dave King has kept his head and got the job done. Well done.

Thanks also go out to her indoors and my little ones, obviously. But mainly I wish to give praise and appreciation to my brave patients, who continue to wallow in the wire of less-than-perfect traditional psychiatric treatments; waiting for the day when I can legally prescribe for them a new range of focused, efficacious and safe adjuncts to psychotherapy. Hold on tight, courageous people, with each passing year that moment creeps ever-so-slightly closer.

CAMERON ADAMS I would like to thank the family that has supported me throughout this venture as a psychedelic public figure: My mom Ann, my brother Gates and his family – Presley, Gates William and Faun. They don't always understand me or what I do, but always support me unconditionally. I would especially like to thank my daughter Zoë. It is for her future that I do this work. Sarah has been a loving sanctuary through some pretty dark moments. I love all of you deeply and unconditionally.

I would also like to thank all of the speakers, artists, performers, volunteers, and curious onlookers for making Breaking Convention what it is. Without you, we'd be five nutters chatting in a pub. With you, this begins to look like the seed of a proper social revolution moving towards a brighter and more brilliantly coloured future. Thank you all for sharing our dream.

AIMEE TOLLAN Firstly I'd like to thank the members and speakers of the Psychedelic Society at the University of Kent, who introduced me to the wonderful world of psychedelic culture and consciousness. Next to the Anthropology department at the University of Kent; the fantastic lecturers and my fellow students who made my time at university transformative, enriching and enlightening. To my family, who have always provided me with a loving and supportive foundation.

The Breaking Convention founders and my colleagues, Ben, Cam, Dave & Dave; thank you for this great opportunity to be a part of the Breaking Convention community. You have been inspirational and I admire you all.

To the rest of the committee; you are a remarkable bunch and I thank you all for bringing your own individualities and skills to make Breaking Convention what it is. Thank you to all the delegates, speakers and other contributors to Breaking Convention for making the conference the special and noteworthy event that it is. Lastly, thank you to all the members of the psychedelic community; thank you for opening my eyes to endless possibilities and giving me hope for positive change.